T0136452

DECADE OF DISASTER

Decade

of

Disaster

Ann Larabee

University of Illinois Press

URBANA AND CHICAGO

© 2000 by the Board of Trustees of
the University of Illinois
All rights reserved
Manufactured in the United States of America
⊚ This book is printed on acid-free paper.

Library of Congress Cataloging-in-Publication Data
Larabee, Ann, 1957–
Decade of disaster / Ann Larabee.
p.   cm.
Includes bibliographical references.
ISBN 978-0-252-02483-2 (cloth : alk. paper)
ISBN 978-0-252-06820-1 (pbk. : alk. paper
1. Disasters—Social aspects. 2. Technology—Social aspects.
3. System failures (Engineering)—Social aspects. I. Title.
HV553.L38   2000
363.34—dc21   99–6114
CIP

1 2 3 4 5 C P 6 5 4 3 2

This place is not a place of honor.
No highly esteemed deed is commemorated here.
Nothing valued is here.
This place is a message and part of a system of messages.
Pay attention to it!
Sending this message was important to us.
We considered ourselves to be a powerful culture.
—Proposed Marker for WIPP
(national nuclear waste site)

Contents

Preface

In the United States, the 1980s have come to be known as the decade of disaster, signaling an age of limits. The technological catastrophes of the decade stimulated debate over the rise of powerful multinationals exporting dangerous systems, the role of modern political institutions in gaining loyalty for massive technological investments, the dangers of an increasingly mechanized lifeworld, the eroding cultural distinctions between humans and machines, and social obligations to disaster victims. In mass media productions, literature and film, philosophy and cultural history, astrophysics and evolutionary biology, popular biography and personal survival narratives, psychology and sociology, and law and public administration, technological disasters were constructed, reconstructed, and subjected to intense examinations. The release of methyl isocyanate (MIC) at Bhopal, the *Exxon Valdez* oil spill, the meltdown at Chernobyl, the *Challenger* space shuttle's fiery crash, the failure of public health agencies to control the HIV/AIDS pandemic, all became anxious spectacles of the failures of modernist technological projects invested in large systems building.

Since 1945, nuclear weapons had defined the limit of high technology, with its massive aggregations of capital, expertise, and hardware. By the 1970s, a full-scale critique of nuclear projects was well under way, with scientists and other concerned citizens publicly imagining worst-case scenarios: accidental launches that escalated into all-out nuclear war, a nuclear winter that froze the earth, and core meltdowns that sent nuclear material all the way through the planet. Furthermore, the technological complexity of life in industrial societies, generated by optimism and complacently accepted during the expansive postwar era, had begun to show signs of disrepair. Old systems were wearing out and new systems were shoddily designed and built, industrial and military secrecy was preventing thorough safety auditing, and pollution was accepted as a normal byproduct of progress and profit. The workplace had become increasingly systemized, with no increase in safety for workers, and was poised for further transformation to complex digital interfaces. When the Three Mile Island nuclear power plant released radioactive steam in 1979, it seemed to prove that a technological disaster involving such large-scale systems was inevitable and that ordinary people would suffer the collateral damage, if they were not already.

Despite these growing fears, the political rhetoric of the Reagan years promoted a vision of U.S. technological progress, competitiveness, dominance, and safety, nationally symbolized in the Strategic Defense Initiative (SDI), or "Star Wars" project. In his March 1983 speech introducing SDI, Reagan called for the nation to "turn to the very strengths in technology that spawned our great industrial base and that have given us the quality of life we enjoy today." This know-how, he argued, would be deployed to prevent an all-out technological disaster with a technological shield, an appealing fiction that held the paternalistic promise of total security from nuclear Armageddon. Thus, technological fixes for complex sociotechnical problems became a matter of national policy, as did a commitment to forming a culture hardened against disaster with networks of satellite surveillance and weaponry, installed under the guise of safety.

Amid these currents of discourse on technological systems, the disasters of the 1980s provoked critiques of corporate and government apparatuses. In keeping with the political climate, they were marked by nuclear images and accompanied by a burgeoning discourse on risk, trauma, and stress. In this book, I explore how these disasters evoked a language of violent disruption, containment, and survival in many venues, contributing to the apocalyptic tone of postmodern cultural formations. Unlike sociological or psychological studies that assume a clearly delineated organizational response to disaster, this book focuses on cultural formations of disaster narratives and their contesting and negotiation.

The writing of disaster often critiques single technological systems but spills out into other cultural venues, including the wider political and social networks in which these systems function. Disasters erupt within various interpretive communities, and in the face of them, the very idea of community becomes contingent, alterable, and even accidental. The destabilizations of the technological lifeworld gain emotive force in the wounded, scarred bodies of the dead and the living. In the explosion of interpretations following the moment of shock, the disaster becomes a critical incident, reflecting on the state of our technological survival.

I have had many guides who helped with my understanding of local issues surrounding disasters. For my research in Alaska, I would like to thank the staff of the Oil Spill Information Center, Martha Vlasoff of the *Exxon Valdez* Oil Spill Trustees Council, Ernie Piper of the Alaska Department of Environmental Conservation, Marilyn Leland and Joe Bridgman of the Prince William Sound Regional Citizens' Advisory Council, Joe Leahy of the Valdez Museum, Mike O'Meara of the Pratt Museum, Jack Hession of the Sierra Club, Dean Glasser of ERA Helicopters, and the locals at the Salty Dawg in Homer. I especially thank Greg Winter, who worked with the Alaska Department of Environmental Conservation during the spill and provided invaluable contact information, documents, and photographs. For my research on Bhopal, Paul Nuchims was very generous in giving information about his community and People Concerned about MIC, the citizen action group in West Virginia's Kanawha Valley. Raine Rönnberg, in Finland, provided me with materials on Chernobyl. My work with the staff, volunteers, and clients of the Lansing Area AIDS Network helped sensitize me to the feelings and concerns of people with HIV and gave me broader understanding of personal disaster narratives.

I also benefited from other scholars who supported my work from its beginnings in 1990. My participation in a National Endowment for the Humanities institute on Rethinking Technology, conducted by Carl Mitcham and Leonard Waks, introduced me to science, technology, and society studies. There, I was lucky to converse with Don Ihde, David Strong, Albert Borgmann, Kristin Shrader-Frechette, Langdon Winners, and other presenters and participants who still influence my writing. I was given a substantial grant by the Critical Incident Analysis Group at Michigan State University to conduct my Alaska research. My department, American Thought and Language, also encouraged me with travel grants and intellectual support. I would especially like to thank chairperson Doug Noverr and associate dean Patrick McConeghy for their steady encouragement, and my colleagues Maria Bruno, Trish Morse, Dean Rehberger, Phil Bellfy, and Ned Watts for reading portions of the manuscript. I am also grateful for the collegiality and commentary of scholars outside Michigan State University, including Steve Biel, Carl Smith, Suzanne Poirier, Tom Birkland, Richard Howell, Nick Cull, Alasdair Spark, and Eyal Amiran. My friends Charley Stivale, Noel Grey, Tersh Palmer, Pranab Kumar, Paul Travers, Melissa Hilbish, Steve Grant, and Denise Pilato were willing to listen to my brainstorming and offered helpful advice. My children, Noah and Lissa Blon, had the patience and good humor to put up with my long hours of singular concentration on a dark subject. And finally, Kathy Sotol first encouraged me to investigate nuclearism and its history, and taught me my first lessons about writing to survive.

An earlier version of chapter 1 appeared as "'Remembering the Shuttle, Forgetting the Loom': Interpreting the *Challenger* Disaster" in *Postmodern Culture Journal* 4 (May 1994) (available on-line through Project Muse). An earlier version of chapter 3 appeared as "Radioactive Body Politics: AIDS as Nuclear Text" in *Literature and Medicine* 13 (Fall 1994): 229–42.

Introduction:
Critical Incidents

At 11:02 A.M. on 9 August 1945, Hayashi Kyoko, a Nagasaki school-girl working at a munitions plant, her hair braided to avoid getting caught in a machine, heard someone cry, "Air raid!" and found herself suddenly blinded by the Bomb. She was saved, she speculated, because she was in a wooden shed, a "beggar's hut," cobbled together from recycled scraps of glass and wood. Her friends who worked in a brick building burned to death, and others were penetrated with glass shards from huge windows in a more modern building. Of that terrible moment of fission, in that "breath of time," she writes, "73,889 people died instantly. Approximately the same number, 74,909 were hurled out in the blazing summer sun, stripped of their skin, like the legendary white rabbit of Inaba who was flayed alive by the croco-dile."[1] At some distance away, her mother attempted to piece together what had happened from falling strips of flowered kimono cloth, newspapers, a wooden picture frame with the word *Nagasaki* painted on it, and a black rain that witnesses were sure was human blood. For Hayashi Kyoko, survival as *hibakusha* had begun in the work of mourning, the remembering of the victims' names, written on long rolls of paper in the school auditorium.

On that same day, as he duly recorded in his notebook, the science reporter for the *New York Times,* William Laurence, flew over Nagasaki in an instrument plane piloted by a philosophy major from the University of Chicago. Listening to the engine, Laurence wrote, "There comes a point where space also swallows time and one lives through eternal moments filled with an oppressive loneliness, as though all life had suddenly vanished from the earth and you are the only one left, a lone survivor traveling endlessly through interplanetary space."[2] And yet he carefully kept a chronological record of the flight for posterity, with frequent reference to actual times. From the pressurized cabin, surrounded by instruments, Laurence looked down on Nagasaki through doubly protective layers: a plastic window and arc welder's goggles. At 12:01 P.M. by his watch, he recorded, the weapon was released from the bomber ahead: *The Great Artiste.* He observed the ball of fire, the "birth of a new species of being" in an accelerated evolution "covering millions of years in terms of seconds."[3] From the technologically enclosed space of the plane, Laurence watched the collapse of time. But he himself triumphantly survived, as though in Einstein's spaceship on the lip of a black hole, in a temporal warp that seemed like a godlike infinity. Such was the view from the air.

In an essay that helped establish nuclear criticism,[4] published in a special issue of *Diacritics* in 1984, Jacques Derrida writes, "If, according to a structuring hypothesis, a fantasy or phantasm, nuclear war is equivalent to the total destruction of the archive, if not of the human habitat, it becomes the absolute referent, the horizon and the condition of all the others."[5] If an all-out nuclear holocaust were to occur, it would erase the production of texts and the entire symbolic realm. Writing therefore must contain the absence of writing, the end of naming, the nuclear destruction of any possibility of writing. Writing is the "movement of survival" that Derrida associates with social memory, culture, tradition, and the symbolic. Any catastrophe, including individual death, that falls short of total annihilation of this archive leaves "the remainder, the work of the remainder," and the work of survival, of writing, is "memory, compensation, internalization, idealization, displacement, and so on."[6] Similarly, nuclear critic Peter Schwenger finds at Ground Zero a "radical instability of language" that might give hope of survival in the face of annihilation.[7]

As the Bomb's "remainders," Kyoko and Laurence both saw themselves as survivors of the modern apocalypse.[8] Their stories intersected in significant ways beyond the fact of the event: the attention to temporal disruption, the sudden isolation, the explanations for why they survived, the attention to surrounding artifacts, the sense of being witness to an historic moment. But there were obviously many more differences in the writers' life histories, drawn along the potent lines of wealth, social position, access to information, and embodiment of technological violence. In disaster narratives, proximity to Ground Zero carries an emotive power. It matters whether witnesses are in the air or on the ground, whether their technological surroundings hold them in safety or suddenly explode into fragments.

In *Technology and the Lifeworld*, Don Ihde extends his discussion of perceptual ambiguities (especially in seeing) to "technology-culture gestalts," arguing that "technologies become variantly embedded: the 'same' technology in another cultural context becomes quite a 'different' technology."[9] Technologies are structured but ambiguous, with multiple dimensions and possibilities, and their cultural reception and adaptation must be negotiated within already existing modes of living. These cultural multistabilities transform the observer. According to Ihde, there is no naive or neutral viewing. I have filtered the stories of Laurence and Kyoko through my own transforming lens to open the examination of multistabilities in accounts of technological disasters. In such narratives, the embedding of new technologies becomes a traumatic incursion rather than a smooth transfer, and old technologies are thrown into a new light of self-destruction. The writing of disaster attempts to stabilize the transforming world.

I prefer the concept of multistabilities to the stakeholders and legal entities of many disaster analysts,[10] who base their model solely on the limited politics of disaster rather than on the wider cultural understandings in which these events unfold. These analysts study various groups who have economic or political stakes in the way the disaster is managed, through litigation or government regulation. Although such studies are valuable, they are limited to a notion of persons as stakeholders having a limited interest in very directed, externally managed outcomes, such as legal settlements, rather than as interpreters who must reinvent life contexts and renegotiate biotechnical aggregations disrupted by disaster. Furthermore, the stakeholder theory suggests that these groups have shared interests that predate the disaster and thus influence their response in a linear path from one condition to the next. This idea is limited because it can't account for ad hoc coalitions, idiosyncratic responses, or other radical transformations in the survivors' states of being, including disruptions in community and language.

Normal Accidents

Perception of disaster is influenced by social narratives of harmoniously embedded technological relations, from a national discourse on safety to the intimate negotiations between people and their household artifacts. In maximalist[11] technological cultures, experience is mediated within complex assemblages of organic bodies, artifacts, and instruments, of completely socialized human and nonhuman subjects "swapping properties."[12] People participating in this collective must have faith in secure, nonviolent human-technology relations. An idealized technosphere is supposed to ensure the complete safety of its inhabitants. In the late twentieth century, within what Albert Borgmann calls the "device paradigm,"[13] complex technologies have a hidden depth, not only in the mystery of their operation, but also in the networks of power that energize them. At the surface, these systems seem to run smoothly, beneath conscious awareness, as if on automatic pilot. Ideally, the human-technology interface is a transparent and harmonious feedback loop, but it often proves abrasive, less than ergonomic, and even life-threatening. Minor accidents recede in the normal operations of everyday life and are tacitly accepted, but a major catastrophe explodes the map and brings a new resolution to systemic networks of power, revealed as fundamentally disaster-ridden.

Disaster narratives serve to mediate, inflate, and assimilate the violence of complex aggregations. Personal survival accounts often evoke a narrator in very ordinary circumstances who suddenly experiences a violent, inexplicable implosion of familiar technological structures. In his classic *Hiroshima*, still used in U.S. high school classrooms, John Hersey describes how ordinary citizens—a clerk sitting at her desk, a doctor reading a newspaper, a priest unloading a cart—abruptly lose their temporal and spatial moorings. Hersey writes that Masakazu Fujii, sitting in his underwear, was suddenly thrown into the river with the rest of his hospital surroundings: "He was buffeted and gripped; he lost track of everything because things were so speeded up."[14] Then he found himself wedged between boards, "like a morsel suspended between two chopsticks."[15] Hersey describes how Toshiko Sasaki, a clerk, was "fixed still in her chair for a long moment," then was buried by the falling ceiling and bookcases holding the business's library: "There in the tin factory, in the first moment of the atomic age, a human being was crushed by books."[16] Through the evocation of the commonplace, these stories imply the secret violence of everyday technologies (chopsticks, books, windows, bricks, tile). A bomb may not only come down and blow the world apart, collapsing time and space, but also implicates other technologies in its work of destruction. Furthermore, it introduces a hidden danger of radioactive fallout, particles that wait in the grass, lie in the lungs, or transect cells, inflicting a damage that unfolds through a lifetime.

Nuclear literature often featured the accounts of the *hibakusha* as both horrifically foreign and alarmingly, even numbingly close, a probable future for anyone living under the nuclear threat. Furthermore, as Thomas Hughes points out in *American Genesis,* the twentieth-century rise of large-scale production culminated in the Manhattan Project, whose unprecedented linking of government, military, university, and corporate entities became a model for other massive technological systems.[17] Beginning with the Three Mile Island accident in 1979, the apocalyptic nature of these systems left the realm of fabulation and became localized and intimate. The experience of the *hibakusha* was no longer exotic, and might even unfold in Harrisburg or Kansas City. The poisonous wind in the household cracks, radiation represented an all too familiar disaster that seemed to underlie all interactions in a biosphere being transformed into a technosphere, characterized by complex assemblies that functioned in local exchanges with owners, workers, and consumers.

In his influential book *Normal Accidents,* published in 1984, Charles Perrow theorizes that these complex systems are inherently and inevitably prone to massive malfunctions. He suggests that uncertainty and error are "normal" in any complex, "tightly coupled" system.[18] In such a system, many components are highly interdependent, so that the failure of one component quickly escalates into disaster. The unfolding of these events can be neither predicted nor prevented. Although Perrow carefully frames certain "systems"—nuclear power plants, petrochemical plants, aircraft and airways, genetic engineering—he uses metaphors that suggest a wider terrain for normal accidents. In his tale, "a day in the life," he describes "your" apparently familiar encounters with overheated coffeepots, lost keys, bus strikes, and faulty automobile parts, all interacting in unpredictable ways to undermine "your" daily schedule.[19] Intended as a parable to illuminate complex, tightly coupled systems, "a day in the life" implies that normal accidents comprise the very texture of technologically maximalist cultures. The plugged teakettle is more than a metaphor for a nuclear plant's core meltdown; it is a component of the complex, uncertain composition of a highly technological lifeworld.

Thus, the failure of one system, such as a nuclear plant or air traffic control, can be read as a symptom of a vast, apocalyptic, seemingly autonomous culture machine, with its blanketing texture and orchestration of flows. At the end of Don DeLillo's novel *White Noise* (1985), its protagonist, Jack Gladney, stands in the supermarket where the aisles have been rearranged, watching the "agitation and panic" of the shoppers, suddenly wandering without a pattern, searching for familiar items. Having survived an Airborne Toxic Event (a cloud of chemicals leaked from a train derailment), the shoppers go down the wrong aisles, unexpectedly stop, crash into each other, and scrutinize labels—"smeared print, ghost images"—for a "second level of betrayal." These symptoms of disarray are normal even in a familiarly arranged supermarket. "In the end," Gladney observes, "it doesn't matter what they see or think they see. The terminals are equipped with holographic scanners, which decode the binary secret of every item, infallibly. This is the language of waves and radiation, or how the dead speak to the living."[20] Here the supermarket is an ordinary place, thoroughly constructed by technological rationality, enclosed in its own logic, its reading of itself. And yet it continually generates everyday fragmentation, confusion, anxiety, loss, misrecognitions, misreadings, meanderings, and collisions, through the very exercise of its own obtuse, insensate rationality. The technological determinism of *White Noise* dampens and absorbs the potentialities of these ordinary disasters, suggesting that there is no escape or resistance no matter how large or small the gaps.

The recent discourse on technology is saturated with such deterministic visions, yet the very writing of the disaster from the survivor's point of view suggests that even the most ordinary person can outlive the worst that her technological constellation has to offer. In the misreadings, meanderings, misrecognitions, compensations, displacements, and internalizations lie the voice of survival and potentials for adaptation and consolidation.[21] One strategy for stabilizing disaster, especially in political and mass media discourse, involves an attempted reinstigation of faith in technologies with a narrow focus on safety. More lifeboats, better radar, improved containment shields, new backup systems: All these improvements are designed to secure social cooperation. Furthermore, the institutional discourse often uses expert studies to prove that the disaster has somehow actually been productive, a boon to the ecological systems. For victims of disaster, the task is to reconstruct family and community relations as a resistance to an apparently senseless technological violence, perceived as an external threat, driven by outside economic and political forces. The violence inflicted on bodies themselves transforms them into maps that can reveal the physical intimacies and betrayals of technological interactions. Taking advantage of breeches, slippages, and ruptures, these actively politicized negotiations stand against the oppressive autonomy of the apparatus.

Archives

The creation of a disaster archive—the compilation, filing, and sorting of disaster narratives—formalizes and consolidates the work of survival. This book, in itself, is a kind of disaster archive, complete with all the vagaries and gaps, displacements and compensations inherent in such work. In the summer of 1996, I had the opportunity to spend time in an actual physical archive singularly devoted to a technological disaster, the *Exxon Valdez* oil spill. Located in Anchorage, the Oil Spill Information Center was initially set up by the Justice Department to aid in damage assessment and to make public selected scientific, legal, and other relevant documents, particularly for litigants. It eventually fell under the aegis of the State of Alaska's Oil Spill Trustee Council, formed to disperse funds from a $900-million civil settlement against Exxon. The mission of the council is to "efficiently restore the environment injured by the *Exxon Valdez* oil spill to a healthy, productive renowned ecosystem, while taking into account the importance of quality of life and the need for viable opportunities to establish and sustain a reasonable standard of living."[22] This broad mandate for restoration means that the archive strives for nothing less than a comprehensive collection of everything written about the spill, a comprehensive writing of the disaster. It contains scientific and sociological studies, government documents, news coverage, historical accounts, amateur and professional video, Exxon's annual statements to its shareholders and public relations brochures, children's books, novels and plays, collections of poetry, photographs, safety films for cleanup workers, advisory bulletins for communities, press releases, and hearing and trial transcripts. The documents collected therein—whether authored by oil industry CEOs, fishermen in small Alaskan villages, or biologists working with otters—have a shared motive: to ensure that such an event is contained and prevented, whether through systemic changes to oil extraction and shipping or emotional appeals to oil consumers.

Texts and images that reveal the extent of damage and restoration, and their subsequent archiving, stand hopefully against another, perhaps even greater spill. The archive sorts out this flow of discourse, files it, and sets an official stamp on various compilations, summaries, and versions, controlling who is represented and who has access. It is here, in the archive, that containment of the heteroglossic shock is truly exercised. However, the archive does not present one official, hegemonic story, or one truth, about the *Exxon Valdez* oil spill. Instead, there are only points of view, or multistabilities, grouped and assigned in a theater of legal operations. With billions of dollars at stake in disaster litigation, all stories have a potential monetary value that gives them dramatic effect in the search for cause and blame and assignments of reparation. Returning to an original moment of the event, outside of these cultural negotiations of meaning, is impossible.

Of course, many writings are not contained in the archive, voices unheard and damage unseen. As Richard Klein suggests in his discussion of nuclear criticism, "The institution of the archive not only makes possible positive remembering; it also permits systematic forgetting—all the possibilities for wandering, for error and discovery, for allusion and influence, for censorship and its undoing, that arise from the intertextual organization of the archive."[23] Furthermore, he suggests that the archive be perceived not only as text but context, including "all the technologies of publication and dissemination, all the systems of retrieval, cataloguing, bibliography, which have made access to the archive possible."[24] Some archival gaps signal a self-protective secrecy, the purposeful withholding of documents and internal memos that might, for example, shed a negative light on Exxon or the state. What is *not* represented speaks of both exercises of power and the absence of power. The vast majority of persons without access to the archive because of poverty, lack of education, feelings of disenfranchisement, or choice are not represented except in the statistics of sociologists and mental health professionals, and their survival unfolds in deeply personal, socially intimate, hidden ways, outside of claims for compensation. And there are idiosyncratic collections of materials, such as a museum director's thousands of black-and-white photographs on contact sheets in three-ring binders, documenting every tiny physical change to his beloved town, caused by the influx of oil spill cleanup workers. Or a former journalist's four cardboard cartons full of clippings, notes, and employment records, all thrown together and stacked in a closet in a council office. Although this book relies mostly on publicly accessible materials, I do not want to dismiss these more private memorializing, rather than monumentalizing, critical acts.

Through such interpretive strategies, survivors make sense of an essentially meaningless violence. As Maurice Blanchot observes in *The Writing of the Disaster*, "The disaster is separate; that which is most separate."[25] It always stands outside our productions of meaning. Survivors of disaster often say that the event was senseless, that they do not know why they were chosen for such an incursion into their lives. The violence is a fissure between a before and an after, an abrupt, exploitable discontinuity. Writing about disaster, collecting and organizing these writings, reasserts the cultural project of signifying, accumulating, and sequencing.

Spectacle

As disaster, with its deadly contaminations of the ordinary world, seems to make up the very texture of modern life, the production of massive spectacles provides public structures for negotiating such a violent state of affairs and identifying (and often strengthening) the streams of power and influence that shape it. As political scientist Murray Edelman observes, not all hurtful conditions become political spectacles. Only those that serve as "reinforcements of ideologies"[26] become significant in a "spectacle which varies with the social situation of the spectator and serves as a meaning machine: a generator of points of view and therefore of perceptions, anxieties, aspirations, and strategies."[27] In the disaster milieu, technological civilization is on trial as it attempts to heal the systemic breach and restore itself through a figural elimination of all risk. Various cultural interpretations, analyses, and judgments attempt reconstruction of a safe world without slippage, broadly defined.

As Carl Smith points out in his history of three violent events in Chicago in the late 1800s, such an imaginative reordering of society has a long history in the United States, as various disaster narratives have been publicly negotiated and have shaped recollection.[28] However, what especially characterized disaster in the 1980s was the attention to large-scale technological systems, influenced by a political discourse on safety from missile attack. The national imaginary "citadel culture"[29] constructed to prevent a fictional invasion of omnicidal weapons also generated the uneasiness of the popular *Terminator* films in which an SDI system, Skynet, has become self-aware and autonomous and run amuck with the aim of destroying all human life. The large-scale disaster spectacles of the late twentieth century, reaching an apex in the 1980s, have had a special social function in arguing that although the technological lifeworld is unpredictable, dangerous, and even omnicidal, it is ultimately survivable. As malevolent cyborgs stomp across a scorched plain of human skulls in the cultural imaginary, some human beings survive to reinvent history at the end of history. The trajectory of technology may seem to move toward a total automation and instrumentalization, with a subsequent destruction of all life, but the disaster itself disrupts this apocalyptic narrative. Any single technological disaster always falls short of the ultimate disaster, and thus provides an analytical site for exploring ways of survival in a technologically mediated world.

News of disaster comes in many ways—in first-hand experience, shared stories, rumors, jokes, and scholarly analyses and investigations—but takes its most highly visible form in the national media, where most people witness the event as an ideological spectacle. Although understandings of psychological and physical survival expand far beyond the narrow realm of the daily news, the most graphic and lasting representations are shaped here, informing subsequent discourses and folding back into disaster communities themselves. In the national media, catastrophes are constructed within a particular constellation of dramatic effect and formulaic narrative.[30] If nuclear omnicide stands now as the symbolic limit of technorational systems, then news productions of catastrophe also carry out the cultural work of survival in this frame of reference, the frame of Ground Zero. A spot on the Alamogordo Desert where the first atomic bomb exploded, popularized in William Laurence's reportage on the Trinity Test, Ground Zero is so often used as metaphor for any disaster center that it has become a media cliché. The concentric circles laid upon a map, surrounding Ground Zero, provide a geometric, interpretive shape to disaster, suggesting objective analysis from the air. As Peter Schwenger writes, Ground Zero is the point of nothingness where meaning collapses, where "the figurative ground of our conceptual systems disappears . . . swallowed by the yawning zero."[31] In the information age, the news is a highly technical conceptual system, and its commentary on collapsing systems—such as industrial plants and airliners—is self-referring. The news of disaster speaks always of its own survival as a sociotechnical system within a civil defense ideology.

During the cold war, especially in the wake of the Soviet development of the hydrogen bomb and the Cuban Missile Crisis, disaster communications and technologies became intimately tied to civil defense. Through radio and television, the Emergency Broadcast System, with its interruption of regularly scheduled broadcasts, might bring warning of anything from tornadoes to nuclear attack. Total disaster seemed always imminent, represented in a blank screen and an eerie high whistle, the sudden disruption of the flow of consumer images, the limit of the media itself. If disaster was always imminent, then crisis *managers* were necessary. This crisis mode constructed a vision of ordinary people prone to hysteria and panic in the face of an ever-impending apocalypse, in need of authoritative voices to enforce social order. As vehicles of early warning and emergency management, the mass media reinforced state authority and fallout shelter ethics, often seeming a futile gesture against a broader cultural vision of the End.

From an airborne perspective, the spectacle's monumentalizing of disaster makes it appear as though seen through a single lens, from a single perspective that flattens events but promises both safety and apocalyptic revelation. Culminating in the television viewer's ride on the nose cone of a smart bomb down to Ground Zero, the media presents a supermarket of images with Wagner in surround-sound where death unfolds but does not matter. Jean Baudrillard suggests that postmodern cultures are an unfolding End, a hyperreal simulation in which meaningless media-created events "follow upon one another, cancelling each other out in a state of indifference."[32] He suggests that such events, including the fall of the Berlin Wall and the Persian Gulf War, are in fact orchestrated nonevents, entirely virtual, without reference to any real events. They thus become self-referential sites for simulations of simulation and continual confirmations of a protected nonreality: "The question is not whether one is for or against war, but whether one is for or against *the reality of war*."[33] Baudrillard holds out little hope of any movement against this "overdense body"[34] of simulation. However, victims of disaster carry a cultural authority in disaster narratives. Proximity to disaster matters because victims of disaster bear witness and carry the marks of violence. Their suffering, which can never be fully represented, stands outside the simulacrum. Thus, survivors are often silenced by institutions and the media and absorbed into legitimation dramas.

However, because disaster narratives are not fully enclosed by the spectacle, spilling out into other venues, survivors often find other means for reconstructing their own damaged worlds. The writing of disaster moves through rearrangements, alterations, adaptations, contingencies, and unexpected affinities, the traces of the living. Even a national tragedy, such as the *Challenger* space shuttle disaster, is not contained within an endlessly replayed video clip and a presidential speech to the dead. It leaves the supermarket of images and takes up residence in other interpretive venues—official hearings, managerial studies, family and community mourning ceremonies, school curricula, physical reconstructions of the wreckage, eyewitness accounts, and protest literature. To subsume these all into media spectacle dismisses many social formations in which disaster discourse is circulated,[35] including traumatized communities where the work of survival takes place. This book explores containment strategies in many kinds of discourse, from institutional attempts to restore legitimacy, to survivors' accounts that engage the work of mourning and community rebuilding. In all of these, relations among people and their technologies must be renegotiated.

In chapters 1 and 2, I discuss attempted reconstructions of institutional and political authority after the space shuttle *Challenger* disaster and the Chernobyl nuclear power plant meltdown. Beyond the obvious nationalist discourse surrounding both events, I turn to institutional constructions of disaster in sociology, education, engineering, public administration, medicine, and other interpretive communities. The work of institution building, promotion, and defense requires an abstract disaster as justification. Disasters become focusing events for institutions, not only for internal reorganization, redefinition, and agenda-setting, but for ideological solicitations of public approval, support, and cooperation. Thus, the *Challenger* disaster, threatening the national space program with all of its culturally symbolic implications, became a focusing event for space advocates, educators, and academic consultants who used it to promote fail-safe systems management for information workers and to define a utopian role for them in a changing economy based on principles of potentially calamitous scarcity. Occurring within months of the shuttle crash, the Chernobyl nuclear plant meltdown also provided an opportunity for institutional articulations of safety through cultural monuments to nuclear disaster. The area around the plant, at first called the Forbidden Zone, was designated a wilderness preserve, and a booming tourist industry arose, culminating in a set of conventional science and travel narratives that sought to reclaim the disaster as ecologically productive and to distance observers from radiation fears. Chapter 3 turns to the advent of HIV in the gay urban community, perceived as a failure of the public health system. AIDS activists transformed the nuclear discourse perpetuated by the national media and biomedical researchers, involving images of radioactive bodies and apocalyptic urban centers, into an articulation of survival and community. Emerging from military research and development of an engineered, disaster-resistant body, the cyborg also emerged in the 1980s to mitigate biomedical disasters, appearing in AIDS discourse to reclaim medical technologies, and in the popular figure of Stephen Hawking, who mediated both black holes and catastrophic illness.

Chapters 4 and 5 examine two disasters involving transnational corporations: the *Exxon Valdez* oil spill and the chemical release at a Union Carbide pesticide plant in Bhopal, India. Both became national media spectacles of the perils of technological systems, emerging finally as relegitimations of conventional human-technology relations, drawn along lines of race, class, and gender. In both cases, the corporations launched a campaign of image restoration that would result in the formation of a public relations specialty in organizational disaster management. Scientific and technical language was deployed in the service of public relations. However, neither Exxon nor Union Carbide was entirely successful at restoring image because of fierce local battles over the meaning of the event. In Alaska, angry residents of oiled communities presented images of dirty crude lifted from the beaches and constructed their own makeshift cleaning technologies, rejecting the invasion of the Texas white shirts. In India, survivors of the Bhopal disaster, erased in the courts and the media, took to the streets over and over again in protest, making their scarred and weakened bodies visible, at the risk of being arrested and beaten.

As the "age of limits," the 1980s saw the emergence of a discourse on both failed technological systems and the cultural work of surviving within such systems. While risk analysts frame survival in terms of statistical certainties, isolating certain dangers such as smoking, driving under the influence, or being hit by an asteroid, disaster explodes unpredictably in a nexus of interpretive possibilities. It is disaster's resistance to any single narrative and its power to force cultural reformations that I explore in the following chapters.

1. Lifeboat Ethics

Lifeboat

In the wake of the World Trade Center bombing in 1993, a made-for-TV, fiftieth-anniversary remake of Alfred Hitchcock's *Lifeboat* appeared, transferring the plot of survival in extreme conditions from ocean to space. *Lifepod* depicts brutal, claustrophobic conditions in a small spacecraft containing a handful of survivors from the terrorist bombing of a much larger space transport. Looking very much like the ocean liner from one of the first large-scale disaster films, *The Poseidon Adventure* (1972), and evoking the recently rediscovered *Titanic*, the large space transport holds too few lifepods for its passengers, and the one vessel that does escape is in bad repair and not sufficiently stocked with food or water.[1] Furthermore, its design is inadequate for space navigation, and its pilot, trapped in a small chamber without solar shields, dies a slow and gruesomely pustulant death from radiation bombardment. The rest of the survivors fight with each other and their depleted technological surroundings until only two remain to be saved.

Blaming the survivors' harsh conditions on a vaguely belligerent, self-serving, and inefficient government authority, *Lifepod* contains a lesson about
preparedness, with traces of the cold war's abandoned fallout shelters and
the crashing *Challenger* space shuttle, which, as NASA critics had often pointed out, carried no escape vehicles. *Lifepod* presents a hostile technosystem
that controls air, food, and water, as the survivors pant, sweat, bleed, freeze,
and starve, at the mercy of their drifting enclosure. This psychologically tense,
physically urgent, claustrophobic existence throws body-technology relationships into sharp relief, but the film argues that rational preparedness is possible if only the government, the shipbuilders, and the owners take responsibility. Although violence may be inevitable, striking at any moment from
some hostile outside force (an iceberg, a terrorist, a tsunami, an interconti-
nental ballistic missile), smooth technological existence might be managed
in a properly maintained, enclosed, and redundant system. Unlike the ne-
glected lifepod, a well-designed, well-stocked escape vehicle would support
its inhabitants in the lifestyle to which they have become accustomed, albeit
more modest, rationed, and rational.

A symbol of preparedness and accurate prediction, the lifeboat is both a
physical and a psychological escape from technocultural terrors and, more
ambiguously, a condensed version of that same technoculture. The lifeboat,
and its more recent manifestation, the fallout shelter, are institutional answers
to disaster: scaled-down models of institutional systems in which persons are
rendered beholden to an authority that may be exercised in severe ways. Of
course, the lifeboat is a smaller version of the ocean liner, without the amen-
ities, where rationing of food and water determines physical and psychological survival. The public fallout shelter, in the civil defense imagination, repli-
cates the disciplining of bodies in school, the public welfare office, the hospital,
and the insane asylum. These exercises of institutional authority, both real and
imaginary, are justified under the improbable conviction that without them,
people will panic and throw themselves or others out of the boat.[2] Furthermore, the lifeboat's limited space makes it a condensed, highly contested mi-
croworld where only a few survivors have access to life support. In this institutional microcosm, lifeboat ethics requires decisions about who survives and
who doesn't, reinforcing social and national distinctions based on race, ethnicity, gender, and class.[3] Triage, which divides persons on the basis of injuries, is also exercised for various reasons, either to ensure speedy medical treatment or to abandon the hopeless. In any event, survivors in the lifeboat or the
fallout shelter must willingly participate in work of the institution under the
threat of death because outside all is radioactive air, deep space, or saltwater.
In *Discipline and Punish,* Michel Foucault uses the medieval plague town as a

metaphor for this fantastical state project of control: "The plague-stricken town, traversed throughout with hierarchy, surveillance, observation, writing; the town immobilized by the functioning of an extensive power that bears in a distinct way over all individual bodies—this is the utopia of the perfectly governed city."[4] Even more than a plague town, with its ever-present, invasive threat of destruction, the lifeboat presents concentrated power as essentially beneficent, the only solution to disaster, and therefore worthy of cooperation. And exercises of power are justified by the limited technological structure that determines survival, a determinism not possible in more open, expansive ecologies.

In a radioactive, polluted, terrorist, and generally chaotic world, one must move to a smaller, safer box, ideally the enclosed world of the harmoniously functioning, disaster-resistant enclosure, a Biosphere 2 that will ensure extraplanetary human survival. The spaceship is a grander version of the lifeboat, designed to move the postindustrial knowledge class away from earthly disasters and into biotechnical systems thinking. Space advocates have long promoted the idea of space travel as a hedge against disaster, most extravagantly the threat of a dying sun: "Remember, the sun blows up in ten billion years, so it's quite clear—we either blow up with it, or it blows up with human descendants remaining alive somewhere beyond the solar system."[5] But the disaster has also been constructed more locally as a nuclear holocaust or the population "explosion": "It is possible . . . to list many excellent practical reasons why Mankind ought to conquer space, and the release of atomic power has added a new urgency to some of these. Moreover, the physical resources of the planet are limited."[6] Because the spaceship has even more limited resources and drifts through a hostile environment replete with cosmic radiation, it is a scaled-down rendering of disaster culture, in which safety and survival provide the overriding logic and in which survivors are selected on the basis of professional skill. These survivors not only are subjected to, but participate in, the disciplinary work of surveillance and observation that seemingly keeps the lifeboat safe, a work possible only in such a microenvironment.

However, the spaceship/lifeboat began to look dangerous indeed when the *Challenger* space shuttle crashed in 1986, after being heralded as "one of the most significant technological achievements of the century" and the solution to human survival in a rejuvenated space program.[7] Critics of NASA asked why there were no lifeboats for the lifeboat, no means of escaping a largely untested, inevitably disastrous technology made up of over 700 critical components, any one of which might cause a fatal accident. Although space enthusiasts may promote the spaceship as a lifeboat in itself, a way of escaping a doomed planet and sowing the seeds of *homo sapiens* across the universe, the shuttle disaster demonstrated that the lifeboat is not especially life-sustaining. Like the unfortunate inhabitants of the negligent *Lifepod*, the *Challenger* seven lived to experience a gruesome drift, the long descending spiral to the ocean where impact crushed the crew cabin. One of the media lessons of the *Challenger* disaster was that in high-tech enclosures, catastrophe is inescapable and its victims—even friendly teachers—have no viable means of ejection. This televised spectacle of claustrophobia and futility riveted millions, who helplessly viewed the exploding microcosm of postindustrial life. Gregory Whitehead writes that the media's construction of the *Challenger* disaster was a "thanaturgical excess of fire + fire + light," a futurist's necrodrama provoking dread and shock.[8]

The failure of this most overtly symbolic national project forced reassessments of the hedges against an always impending disaster: containment, safety, and escape. Assessing the shuttle's future, NASA and its advisers deemed rescue vehicles too heavy, dangerous, and impractical to be incorporated into future shuttle designs. Ironically, the institution of such safety systems was thought to make complex technologies even more dangerous and unpredictable. Nevertheless, in the disaster's self-critical wake and anticipating the use of the Soyuz capsule as a lifeboat on the accident-ridden *Mir* station, Jerry Craig, head of NASA's Crew Escape and Reentry Vehicle planning office, suggested that the planned U.S. space station be provided with enough lifeboats for everyone. In answer to critics who felt that the space station should be made safe enough to do without lifeboats, Craig writes, "That's kind of like saying the *Titanic* would never sink."[9]

Because the *Titanic* wreck had been discovered only a few months before, the calamitous failure of the shuttle was inevitably compared to the sinking of the great ship. Because of its recent press, the *Titanic* was repeatedly evoked as the *Challenger*'s unheeded precursor, another ship doomed by ice, faulty design, components adversely affected by cold temperatures, and official mismanagement.[10] The *Titanic* legend's principle themes—"man" against "nature," reckless speed, technological hubris, and lack of adequate lifeboats—all fit easily into the shuttle disaster's postmortem. But because Robert Ballard had found the *Titanic* wreck using robots, submersibles, and satellite navigation systems, this new narrative of technological control over a historic disaster could also be mapped onto the *Challenger*.

The *Challenger* salvage allowed for a public demonstration of a restored technological mastery through surveillance and recovery. Recovery of the wreckage, strewn over some 420 square miles of ocean, required a massive effort lasting months. As one reporter for the Navy personnel magazine, *All*

Hands, writes, the salvage was "an exciting and professionally rewarding assignment."[11] The technologies that had located the *Titanic* were used to find, photograph, and identify the space shuttle's hundreds of scattered parts, of great interest to the press and the presidential commission investigating the accident, headed by former Secretary of State William Rogers. As the head of the search, recovery, and reconstruction team reported to the Rogers Commission, the manned submersible *Sea Link II* was one of the most successful vessels in locating the shuttle's debris. Because the *Challenger* disaster had fueled the debate over the cost, safety, and viability of manned space flight, the operation of a deep-diving manned submersible in another hostile environment was a vindication, its success duly entered in the official testimony of the Rogers Commission report. In its post-*Challenger* report to the president on the space program's future, the National Commission on Space called for a small orbital maneuvering vehicle, an "*Alvin* Submarine for Space," referring to the submersible popularized by the *Titanic*'s discovery.[12] The deployment of successful submersibles and the reconstruction of the *Challenger* were political symbols of a resurrected faith in national technological projects that had to be proven safe.

The Rogers Commission investigation of the accident attempted to restore this faith in the safety of all human-technology relations, not only in high-tech microenvironments such as submersibles and shuttles, but in the more familiar technosphere of daily life. The Rogers Commission determined that the shuttle crashed because of the hot gas breach of a seal, an O-ring. In an effort to make this understandable to the public, government officials and journalists made these faulty parts seem as simple as putty and rubber washers, familiar to anyone with a leaky faucet. In her thorough investigation of the organizational communications leading to the shuttle disaster, sociologist Diane Vaughan describes her own difficulties in understanding the technical details of the leaky joint, first imagining it to be a Nerf ball, and then a "rubber ring between a Mason jar and its lid."[13] The booster seal was the central focus of testimony from engineers, who described evidence from earlier shuttle flights of blow-by—the leaking of hot gases from the booster seals. Blow-by was indicated by the presence of soot, ranging in color from gray to black. According to Morton Thiokol engineer and whistle-blower Roger Boisjoly, black, which appeared when the seal was subjected to cold temperatures, indicated that it was going "away from the direction of goodness."[14] When the *Challenger* was launched under cold temperatures on the morning of January 28, the seal failed completely and the shuttle caught fire. The Rogers Commission verified suspicions that the poorly designed seal of the right solid rocket booster was the technical cause of the accident. But it also accused the managers of NASA and its contractor for the solid rocket boosters, Morton Thiokol, of not heeding early warnings from engineers about the faulty seals.

Consisting of five published volumes, including 1,700 pages of testimony and numerous appendices containing charts, graphs, and parts lists, the Rogers Commission report resembles product liability trials that set out to identify the responsibility for the technological failures of daily life—faulty wiring, exploding gas tanks, toys small enough to choke infants. According to Elaine Scarry, the product liability trial is a "*cultural self-dramatization:* The courtroom is a communal arena in which civilization's ongoing expectations about objects are overtly (and sometimes noisily) announced."[15] Here, a narrative of disaster is constructed in order to restore civilization: "Implicit in this mimesis of restorability is the belief that catastrophes are themselves (not simply narratively but actually) reconstructable, the belief that the world can exist, usually does exist, should in this instance have existed, and may in this instance be 'remakable' to exist, without . . . slippage."[16] Part of this remaking is enacted through compensation for bodily damage, a healing of technological wounds through judgment and financial reward.

Like the judge and jury in a product liability case, the Rogers Commission was certainly engaged in a remaking of civilization and its projects. Revisionist risk theorists see disasters as opportunities for reconstructing dangerous institutions through a minute investigation: "Each piece of physical evidence . . . becomes a kind of fetish object, painstakingly located, mapped, tagged, and analyzed, with findings submitted to boards of inquiry."[17] But there is a larger necessity embedded here: regaining public trust in big technological projects that justify the state. The trial was enacted before the public eye, a national demonstration to restore the narrative of technological progress with testimony from scientific experts. The commission's broad mandate was to "review the circumstances surrounding the accident" and "develop recommendations for corrective or other action."[18] And this mandate was framed by a "firm national resolve"[19] to restore the space program—a program that has reified national identity around a supposedly common endeavor that transcends cultural differences, preservation of the species from wholesale disaster. In the many reiterations of the steps that led to disaster, in the meticulous documentation of the shuttle components' performance and NASA decision-making hierarchies, the Rogers Commission report sought to reinvent the nation—and indeed all human making—without blow-by and slippage.

The most spectacular moment in the Rogers Commission's testimony was when commission member and eminent physicist Richard Feynman dropped a bit of O-ring material into a glass of ice water to prove its lack of resiliency under cold temperatures. Immediately picked up by the press, who lionized Feynman, this simple impromptu experiment seemed to cut through the waffling, confusing, jargon-riddled rhetoric of the NASA decision-makers' testimony. But perhaps more importantly, the experiment demonstrated that catastrophic failure occurs in basic technological parts and everyday household experience. As engineer Roger Boisjoly later claimed, "most failures usually occur because some minor subsystem gives: 25-cent washers, $2.50 bolts, $25 clevis pins."[20] The press claimed whistle-blowers Feynman and Boisjoly as heroes precisely because they seemed to expose the simple truth about quotidian life in the technological age. Our most familiar objects carry incipient, unforeseen, body-threatening dangers. In his discussion of technological accidents, sociologist Ron Westrum writes, "A computer chip smaller than a thumbtack can send an airliner crashing into a hillside."[21] The preface to the Rogers Commission report states, "The Commission construed its mandate somewhat broadly to include recommendations on safety matters not necessarily involved in this accident but which require attention to make future flights safer. Careful attention was given to concerns expressed by astronauts because the Space Shuttle program will only succeed if the highly qualified men and women who fly the Shuttle have confidence in the system."[22] As a public hearing on body-technology relations, the commission's report attempted to restore confidence in even minor subsystems, to reinstate a national faith in technological existence, made safe through vigilance and the most minute surveillance, down to the thumbtacks.

Disappearing Bodies

What is most strikingly absent from the remade world of the technocractic Rogers Commission report is any effort to reconstruct and assess bodily damage. Although it opens with the now famous photograph of the smiling shuttle astronauts and payload specialists in their shiny sky-blue space suits, posed with an American flag and a toy model of the *Challenger,* the report contains no discussion of the bodies. Often working on their hands and knees in low-visibility conditions, salvage divers found the corpses several weeks after the accident. After a civilian diver spotted a space suit, the Navy team discovered the crew compartment where they "could read the nametags on the astronauts' blue flight suits."[23] The Rogers Commission took testimony until early May but received almost no forensic evidence, nor did the commission express any desire for such evidence, stating publicly that this would be inappropriate and outside its jurisdiction.[24] The only exception lies in the 7 February testimony of FBI special agent Stanley Klein, who reported, "We do have human hair, Negro hair, Oriental hair, and hair from two different brown-haired Caucasians, and what is interesting, according to the laboratory, is that there were no signs of heat damage to any of the hair, which was surprising. The hair came from face seals, fragments of helmets, and helmet liners, and headrests."[25] This reduction to anonymity of NASA's highly touted racially and ethnically diverse shuttle crew was quickly passed over in favor of a discussion of possible laser terrorism by Libyan dissidents and Puerto Rican proindependence groups.

The Rogers Commission followed NASA's lead. NASA's position in the disaster's aftermath was that the astronauts and payload specialists had died instantly, or at least were rendered instantly unconscious, an assumption easily accepted by television viewers who had watched the fiery crash. However, careful study of footage from the disaster clearly revealed that the crew compartment had hurtled to the ocean intact during a nearly three-minute descent, and NASA later revealed that some crew members had activated and

Challenger space shuttle crew. (Courtesy of NASA)

used their emergency air packs. Controversy arose over NASA's original transcript of the voice recorder that elided pilot Michael Smith's ominous, "Uh-oh." Attempting to restore public faith in technology, neither NASA nor the Rogers Commission was very willing to admit that the crew might have known its fate and suffered.

The strict control of information surrounding the bodies of the lost *Challenger* astronauts and payload specialists had purposes beyond delicacy and respect for the crew's loved ones. Their long and horrifying deaths had to be suppressed in the interests of continuing manned space flight. With eminent astronauts Sally Ride and Neil Armstrong participating, the presidential investigative committee remained committed to manned space flight, hearing from other astronauts such as former *Challenger* pilot Paul Weitz, who testified that "man can do many wonderful things in orbit."[26] However, Weitz also suggested, "Every time you get people inside and around the orbiter you stand a chance of inadvertent damage of whatever type, whether you leave a tool behind or whether you, without knowing it, step on a wire bundle or a tube or something along those lines."[27] Although the enormously complicated technologies of the space shuttle might, in ideal circumstances, provide a secure enclosure for experimental human and animal bodies, those bodies are marked by mundane clumsiness, inadvertent behaviors, everyday chance, and uncertainty.

Furthermore, bodies are not especially suited for life in space. On long flights they are subject to muscle and bone deterioration and weight loss, and ubiquitous radiation may damage reproductive organs. As NASA consultant Harry Shipman explains in his book about the future of space flight after the *Challenger* accident, bodies pollute spacecraft, contaminating them with sweat and transforming them into smelly "urine dumps."[28] Whereas male astronauts in the good old days used catheters and plastic bags, the presence of women necessitates more elaborate plumbing; the shuttle's zero-gravity toilet, the "slinger," caused "serious problems in actual use and . . . required a good bit of cleaning."[29] During the May 1985 flight of the *Challenger*, twenty-four rats and two squirrel monkeys being tested for their responses to weightlessness produced an unanticipated "flood" of feces, so that the uncomfortable crew had to wear face masks.[30] The scatological body, especially the female or animal body, mars the strictly hygienic myth of the clean machine. A dead body is even worse.

The fundamental question in the decades-long argument over manned space flight is whether bodies need to be present at all. As the eminent physicist James Van Allen wrote in the wake of the *Challenger* disaster, "all the truly important utilitarian and scientific achievements of our space program have been made by instrumented, unmanned spacecraft controlled remotely by radio command from stations on the earth."[31] Furthermore, new virtual reality technologies might even project human senses into hostile environments. If humans on Earth can operate finely sensitive space robot arms and eyes or drift remotely through hallucinatory worlds more fantastic than alien planets, why are their bodies necessary in space? Recommending entirely automated extraplanetary operations for a renewed space program, artificial intelligence expert Marvin Minsky writes, "As for safety, *no one gets injured when no one is there.*"[32] The loss of the *Challenger* seven called into question NASA's commitment to the "man-machine mode" in space travel and made its defense of a human presence in space even more controversial. In its 1985 study *The Human Role in Space* (THURIS), NASA laid out its theory of cybernetics, insisting on the presence of humans in a largely autonomous system: "There is no such thing as an unmanned space system: everything that is created by the system designer involves man in one context or another; everything in our human existence is done by, for, or against man. The point at issue is to establish in every system context the optimal role of each man-machine component."[33] THURIS created a taxonomy for human-machine interactions: manual (hand tools), supported (manned maneuvering units), augmented (power tools, microscopes), teleoperated (remote control systems), supervised (computer functions with human supervision), and independent (artificial intelligence). These categories do not make much sense in themselves; clearly some manual manipulation is required for power tools, and microscopes, wrenches, and hammers augment and support human capabilities. But the taxonomy inscribes a fossil record, a technological evolution toward "self-actuating," "self-healing," independent machines.[34] The THURIS authors hoped that such independent machines would require human intervention and attempted to describe uniquely human contributions to largely automated space enterprises. Humans, they argued, possess the unique capacity for visual evaluation, motor coordination appropriate to complex assembly, and mental powers of interpretation, innovation, deduction, and judgment. (Recent developments of artificial neural networks and fuzzy logic call even these "human" powers into question.) According to THURIS, the least important aspect of human intellect is memory: "Man's memory, of all intellectual capabilities, is the one most easily duplicated and surpassed by computer activities."[35] Memory, the basis of culture, becomes unnecessary when human workers function within the machine.

This discussion of what makes a human presence necessary to space technologies reflected broader cultural anxieties about the changing nature of work, an apparent shift to an "informated" postindustrial society, identified by work historian Shoshana Zuboff in *In the Age of the Smart Machine* (1984). Zuboff argues that new computer technologies have destabilized traditional work hierarchies and erased embodied techniques. Unlike automating, the "informating" of production not only replaces the laboring body, but puts production under continuous surveillance, redefines skill as symbol manipulation, and redeploys the body as a social instrument for organizational communication.[36] Zuboff proposes two possible scenarios for the future of the informated workplace: one in which managers reconstruct their authority by emphasizing a panoptic, "fail-safe" machinic intelligence over worker knowledge, and one in which workers are encouraged to use new "critical judgment" in "understanding and manipulating" instrumentalized information, resulting in a blurring of traditional power relations and greater collaboration.[37] Clearly promoting the second alternative, Zuboff argues for a "fertile interdependence between the human mind" and its machines and a "comprehensive vision" of harmonious domestic relations and individual fulfillment made possible through this interdependence.[38] Like the THURIS authors and many other critics of artificial intelligence in the 1980s, Zuboff defends human contributions, unparalleled by the machine: "The informated organization . . . relies on the human capacities for teaching and learning, criticism and insight."[39] *In the Age of the Smart Machine* articulated a new technorational, utopian organization for fertile knowledge workers, decontextualized from other ecologies and social relations.

This dream of a productive informated worker, unique and essential to the corporate machine, took a nationally symbolic shape in the promotion of the space program, always a public forum for elites invested in high technologies. The knowledge worker, proficient at operating, interpreting, and mediating new technologies, replaced the cold war's spaceman, who displayed physical endurance and psychological stamina within the hostile technosystem of the claustrophobic capsule. Thus, the *Challenger* seven crew consisted of a social studies teacher, an electrical engineer, a physicist, and a corporate representative from the Hughes Aircraft Company. Physical fitness was unnecessary; journalist Malcolm McConnell observed that Christa McAuliffe was "a little chubby" and that Greg Jarvis "could have easily lost ten or fifteen pounds."[40] The *Challenger* crew represented a populist presence in space: entirely informated workers living happily within the machine, untainted by global political and environmental concerns, whose function was to dispense information to outside observers. Dwarfed by the massive shuttle, their mission was to mediate and domesticate the machine for a young television audience—Christa McAuliffe was to have taken her remote students on a video field trip around the Orbiter. But after the destruction of the harmonious technoworld presided over by a teacher mom, psychologists and grief specialists raced in to erase the spectacle of graphic technological violence and the imagination of Christa McAuliffe's body. In her book on cultural sanctions against women in space, Constance Penley argues that NASA's choice of McAuliffe, in the tradition of "Republican motherhood," served to "domesticate" the agency, and that the privacy surrounding her last moments created a psychological void into which all kinds of public discourse could be projected.[41] Like the children of Bhopal who still reenact the chemical disaster in their games, U.S. children were especially vulnerable to this construction of disaster, imagining car crashes and furnaces and hot water tanks exploding in their houses. They were encouraged to reenact the moment of trauma through videos and toys, sometimes replaying the disaster with a happy ending.[42] In the discourse of the *Challenger* disaster, the corpses of the shuttle crew had to remain behind the technological veil, in the interests of continuing manned space flight and the cultural renegotiation of the necessary body.

However, folklore scholars have noted that the many popular jokes emanating from the *Challenger* disaster often involved those bodies in quite graphic ways. These jokes present the body-technology interface as a spectacularly violent one, as opposed to the cultural ideal in which interaction between the human body and the machine is a flow state:

Q: What do you call a burnt penis on the Florida shore?
A: A shuttlecock.[43]

Q: What was the last thing that went through Christa McAuliffe's head?
A: A piece of fuselage.[44]

Q: Why didn't they put showers on the *Challenger*?
A: Because they knew that everyone would wash up on shore.[45]

Based on familiar rhetorical patterns and cycles, these "sick" jokes have been called political cynicism, a rebellion against the mass media's pompous reverence, a critique of national institutions, and an alleviation of death anxieties in the nuclear age.[46] We expect our technologies to be transparent so that ideally we are scarcely aware of the machine's presence in the cybernetic flow. For example, we expect our telephones to bring us the voices of our loved ones as if they were really present, rather than coded into energy impulses in fiberoptic cables. Skilled operators are supposed to become one with their machines; distinctions between the organic and the technological disappear in harmonious signal and response. Technological disaster shifts the terms of that interaction, for here technology violently entraps, penetrates, and chars the body locked in its embrace. It is this possibility that evokes both national efforts at repression and the return of the repressed through the joke cycle. In a national spectacle of disaster, the body is the pain of technological violence that can never be represented, but only displaced by word and image. Thus, the body must be reconstructed within an organizational safety model, a lifepod, that denies any further possibility of collapse.

Groupthink

The national hearing on the space shuttle disaster attempted to restore the idea of a safe and efficient manned organization, made possible through more fluent exchanges of information. The Rogers Commission report made it clear that NASA's organizational decisions were to blame in the decision to launch the space shuttle, despite icy weather and faulty booster seals. Thus, NASA's management, as well as failed machine parts, became an object of study. NASA's organization was represented in the Rogers Commission report as a self-regulating system without external surveillance or intervention, a situation sociologist Diane Vaughan credited, in part, to NASA's secret military projects.[47] An effective external regulator would have had access to classified materials, an unacceptable risk in the cold war climate. Without external reality checks, many critics suggested, NASA had become isolated in its own delusional can-do ideology, derived from its *Apollo* mission successes. Furthermore, media coverage of the Rogers Commission hearings displayed the homogeneous makeup of NASA administrators and its corporate engineers—all middle-aged white men with a lifelong devotion to NASA and the aerospace industry. Observing the "contradictory" and "rancorous" displays of agency infighting at the hearings, the *New Republic* suggested that NASA itself seemed to be experiencing a midlife crisis.[48] The modern organizational man was exposed and displayed through the figure of the NASA administrator, locked in a decaying airtight compartment of his own making and possessed of the wrong stuff.

In the scientific press, especially in the first assessments of the disaster, some attempt was made to blame NASA's rank and file. A few weeks after the disaster, *Science* twice reported that an internal review of the shuttle had found "relaxed workplace standards" including "worker inexperience, lack of motivation, and faulty equipment."[49] Furthermore, it indicated that NASA's investigation included speculation that workers had forgotten to plug a hole in the faulty booster after a leak test.[50] Despite the search for "inadvertent damage" caused by flawed workers, blame was soon leveled at NASA's and Morton Thiokol's decision-makers, who came to represent a nationwide corrupt power elite, now open to investigation. Charles Perrow, whose study of accidents in complex systems would often be evoked in discussions of the *Challenger*, decried the "Pentagon effect" at NASA that created a climate of managerial self-aggrandizement and toadying to corporate and military sponsors and the media.[51] Journalist and longtime NASA observer Malcolm McConnell wrote that "the rank and file people in NASA are among the hardest working, most productive, and most talented employees in the federal government."[52] McConnell blamed ambitious policymakers engaged in "the political intrigue and compromise, the venality and hidden agendas" that led to disaster.[53] In another account, Joseph Trento also called the disaster a political failure, quoting shuttle mission specialist John Fabian on the *Challenger* investigation: "It just unraveled like Watergate."[54] Thus, discussions of the *Challenger* disaster spread beyond mechanical error to wide critiques of postindustrial capitalists, skilled at political manipulations in a secretive high-tech world.

The national media harkened back to NASA's glory days, suggesting that the organization had devolved, degenerated, decayed from a golden age of right rule, benevolent and safety-conscious. The same space journalists who attacked a highly politicized NASA rhapsodized about the pride and the glory, the "heroic neoclassical élan of the moon race."[55] Little connection was made to NASA's ever-recurring technical failures, including the horrifying *Apollo* space capsule fire that entrapped three screaming astronauts in a fiery furnace and melted them into a nylon puddle. Nor was much mention made of NASA's origins: the "Rocket State" developed in tandem with nuclear weapons and fueled by cold war paranoia.[56] This lack of a thorough cultural critique left a way open to NASA's salvation from a wholesale attack on the disastrous, deviant operations and inevitable risks of complex systems.

The vision of NASA as a once-effective, decadent organization was very appealing to academic theorists who set about to "fix" the agency, using it as a research model. In the flurry of sociological studies that followed the *Challenger* disaster, NASA's homogeneity and in-group ambience, its hidden agendas, political maneuvering, and backstabbing, came to signify the internal workings of all corporations. Social theorists searched for ways to explain and heal the breach in organizational systems, dissected and exposed in a public hearing, fanned by a nationally televised tragedy. Academia, in itself a largely homogeneous entity with its own industrial and military affiliations, responded to the *Challenger* disaster with a corporate consultant's enthusiasm. Discussions of what went wrong with NASA became a common pedagogical tool in public administration, political science, and sociology courses.

Ensconced in university government documents sections, the five-volume, disembodied Rogers Commission report provided an easily accessible text for application of organizational theory and systems models, based on information flow within conveniently closed circuits. According to organizational theorists, NASA was, like the space shuttle itself, a malfunctioning but correctable system with faulty components—namely, NASA's and Morton Thiokol's managers and engineers and NASA's external and internal regulatory units. NASA had experienced blow-by and slippage in its communication linkages: Some of Morton Thiokol's engineers had attempted to voice their fears about the faulty booster seals and cold-temperature launches to their bosses, who had essentially ignored what they considered unproven speculations.

Many theorists attributed the communication failure to NASA's fall from grace. According to this scenario, NASA once had "a less hierarchical and flexible matrix structure" that relied on "nurturing consensus."[57] From these days of childhood innocence, the agency had grown increasingly isolated, streamlined, and pressurized, indulging in overweening bureaupathological fantasies about its abilities, despite budget cuts. In addition, NASA's components had become highly specialized in their activities, languages, and fundamental worldviews so that, for example, the professional ethics of engineers did not match the expedient decisions of managers.[58] Isolated from engineers, NASA's management engaged in groupthink, driven by fantasies of invulnerability and a need for unanimity and cohesion.[59] Thus, the decision to launch the *Challenger* was a technocracy's "major malfunction."

Despite rumblings in the media that the space agency was in its last hours after an apocalyptic failure, academic theorists accepted NASA's continuing existence at face value. Like the shuttle, it was a machine that could be repaired through better interactions and linkages among its components. The machine was wearing out, but it could be restored through an overhaul. Engineers and managers could be realigned. Better brakes could be put on quick decisions. Communications and regulatory valves could be cleaned of soot and debris. The processing system could be repaired to allow the correct flow of information energy, to prevent lacks or excesses of data, to turn away maladaptive codes. Then, tires kicked, the ship would be ready to sail to Mars with human and animal bodies safely enclosed.

The *Challenger* disaster provided organizational theorists with an opportunity to show that the systems model applied equally well to machines and human societies. Using Charles Perrow's work on accidents in complex systems, Diane Vaughan wrote that technological failures could not be separated from organizational failures and that the language of systems applied to both. NASA "malfunctioned" because "the failure of one component interacts with others, triggering a complex set of interactions that can precipitate a technical system accident of catastrophic potential."[60] The use of systems theory in critiques of post-*Challenger* NASA is disputed by G. Richard Holt and Anthony W. Morris, using Yrjö Engeström's activity theory, acknowledging that human activity is not technologically determined, but "messy, disorganized, seemingly chaotic, and hence endlessly fascinating."[61] To ensure safer space flights, Holt and Morris argued, NASA had to accept the internal contradictions and wide possible outcomes inherent in such human activity. Although the authors exposed gaps in systems models of NASA, their aim was to fix the agency as an information-processing system, a contradictory position in itself.

The *Challenger* catastrophe threatened political mythologies of the final frontier and, in a larger sense, cast doubt on systems theories and the entire cultural project of systems building. In his *Evolutionary Systems and Society*, Vilmos Csányi writes that systems models, despite their predictive value, can only approach the "ontological complexity" of nature, but "the interactions of matter . . . are infinite and immeasurable."[62] Thus, the systems model can only represent a semiotic, self-referential complexity. The models of organizational theorists reflected the strict methods of disciplines and vested interests in the national space program. A radical sense of discontinuity, uncertainty, potentiality, and violence—the ontological complexity of disaster—threatened the fundamental order of disciplines, apparatuses, and methods. Charles Perrow put this in the strongest terms reminiscent of the 1960s radical left: "Risky systems are full of failures. Inevitably, though less frequently, these failures will interact in unexpected ways, defeat the safety devices and bring down the system."[63] Thus, the academic response to the *Challenger* disaster was an effort to restore stable systems, and, in an entirely self-referential mode, to reassure its academic audience of information workers that their systems, ideologies, disciplines, and bodies were still in place and all was right with the world. There might yet be a teacher in space.

To the Stars

One of the broader political outcomes of the *Challenger* disaster was a massive public relations campaign by space enthusiasts to resell the idea of manned space flight. The National Commission on Space, appointed by Ronald Reagan, produced a strategic planning report in 1986 on the future of space ventures that included renewed shuttle flights, construction of space station *Freedom*, increased space surveillance of the biosphere, and human settlement on the moon and Mars. In 1989, George Bush called for a lunar settlement by 2004 and a manned trip to Mars by 2019. In 1990, the United Nations endorsed 1992 as International Space Year (ISY), the quincentenary of Columbus's landing, inflaming the usual cant among U.S. politicians and space enthusiasts about human destiny, pioneering spirit, and life on the new frontier.

In that same year, Philip Robert Harris, a "management and space psychologist" and NASA consultant, published *Living and Working in Space: Human Behavior, Culture, and Organization,* an attempt to justify the use of the behavioral sciences in space settlement design, using James Grier Miller's living systems theory. The book is introduced by Jesco von Puttkamer, a NASA program manager and strategic planner, who briefly describes the post-*Challenger* NASA as rejuvenated, ready to "penetrate the new frontier of space."[64] Von Puttkamer argues that the *Challenger* disaster provoked a public outpouring of support for manned space flight because of an "unconscious, unspoken feeling that we are dealing here with evolutionary forces at work."[65]

In behavioral science, evolutionary biology, and artificial intelligence research, systems theory proposes that the biosocial world is made up of systems with interactive components, allowing flows of information and energy.[66] According to these theorists, a natural, intuitive law dictates that systems evolve into more and more complex entities (for example, molecules-cells-organisms-ecosystems-biospheres or cells-organisms-groups-societies-supranational systems). The evolution of Earth systems under the influence of matter, energy, and information flows has resulted in a global, biocultural, technologically regulated supersystem. Thus, the worldwide cybernetic information exchanges of the postindustrial world are seen as the result of thermodynamic, evolutionary processes leading to higher organizational levels.

Systems theorists associated with space programs see human expansion into space as the next organizational level beyond the biosphere. Thus, von Puttkamer writes that manned space travel allows "Man," "Earth," and "Space" to be "one single creative system," an "intricately closed-loop feedback system, a super-ecology."[67] In addition, the formation of extraplanetary artificial biospheres will be designed for what von Puttkamer predicts will be a new cybernetic species, a weightless species, floating in a space womb, transcending gravity and "entropic deterioration."[68] These ideas of evolutionary expansion into space reflect the principle of plenitude, a persistent idea in Western culture that God created life to reproduce richly and diversely and fill the Void. Thus, John Allen, creator of Biosphere 2, the desert amusement park and scientific experiment in space living, explains that his project will expand life's quest to fill all available econiches, hedging its bets against catastrophe.[69]

The idea of an impending catastrophe, most recently by nuclear war, environmental disaster, or cometary collision, is a favored reason for human extraplanetary expansion. During International Space Year, Charles Walker, assistant to the president at McDonnell Douglas and president of the National

Space Society, explained his support of manned space exploration: "Human survival. Political and economic survival in technical competition within the global economy, sure. But more than that: All human creation, all life as we know it, is here on earth. All our eggs are in one basket, one planet. But our embryonic resources are diminishing, and our nest becoming fouled. Our technological nature has given us the means to remove that risk."[70] Here, haunted by the specter of catastrophe, the dreams and aspirations of the postindustrial knowledge class[71] have been given the shape of science fiction and justified through the nineteenth-century language of evolution and nature and the twentieth-century language of systems. The rhetoric of eggs and nests reminds us of the dinosaurs, now popularly recognized as warm-blooded, egg-laying, and nurturing creatures wiped out by a cometary collision that brought nuclear winter to the earth. Frequent evocations of "eggs," "embryos," "cradles," and "wombs" reinscribe sexual reproduction within an entirely mechanical environment, a protective exoskeleton of metal plates that will protect, control, and manage the human body and ensure the genetic continuation of the chosen spacefarers. Ironically, human sexual reproduction in space may actually be impossible, under weightless, radioactive conditions.

In this fantasy of a postindustrial army in space, fears of impending accidents like the *Challenger* make all cultural expression a safety function. Indeed, space planners have invented a culture of catastrophe based on faith in prediction. Catastrophe provides the rationale for subsuming the disciplines under "spaceology," the transformation of the body into a stable energy-matter-information channel, and the continual mapping and surveillance of system biotechnical components. This national vision of the human future counters (and thus depends on) the construction of the thrilling and threatening mass media cyborg, imaged as the Terminator or Robocop, who performs destabilized and penetrated social identities.[72] Furthermore, the national science fiction of space travel seems reassuring next to the spectacles of disaster in the 1980s, not only the real-life disasters of leaking toxic chemicals and exploding machines, but those designed for entertainment: graphic nuclear holocausts with shriveling humans in flames; raging dinosaurs ripping men in half; artificially intelligent computer systems trapping and suffocating workers; buildings exploding and falling into gaps in the earth, crushing their inhabitants; planes crashing in an elegant bloody montage of flying shrapnel. Space planners reassure us that catastrophe is our origin and our nature: The Earth-crossing asteroid or comet that destroyed the dinosaurs "allowed a tiny creature, the ancestral mammal, to grow, differentiate, and fill vacated ecological niches, giving rise eventually to *homo sapiens*."[73] Those asteroids can now be mined for hydrogen, carbon, and nitrogen to feed the transcendent biotechnical organism of the postindustrial knowledge class, emptied of troublesome memory, safe at last.

The political and social meanings of this consensual future are quite apparent in the imagined space settlements of *Living and Working in Space*. Philip Harris refers to the expansion of the human species, the global human family, into the solar system, fulfilling a natural urge for frontier exploration. But his space settlements are built and inhabited by only a segment of that family, the postindustrial knowledge class, envisioned as a cross-disciplinary group of scientists, engineers, technicians, corporate managers, psychologists, sociologists, anthropologists, physicians, teachers, journalists, lawyers, politicians, architects, filmmakers, and designers. Harris writes, "The colonists to the New World during the eighteenth century were largely poor, ill-used white artisans and indentured servants, as well as African slaves. The prospects are that space colonists of the twenty-first century will be more affluent and self-directed, better educated and chosen. Expertise is required of specialists in cross-cultural relocation and living in exotic environments to design systems for deployment and support of spacefarers."[74] Thus, the *Challenger* disaster provided the text for the postcatastrophe survival of the knowledge class, constructed and maintained through systems theories. The *Challenger* disaster suggested that technological and organizational systems were ever on the verge of collapse; the massive public relations campaign for space settlements imagines a safe new artificial biosphere, a closed ecology, for the chosen: academics, civil servants, and corporate managers, freed from environmental disaster, atmospheric impurity, starvation, poverty, disease, and gravity. Harris suggests that this cross-disciplinary community will result in a transformation of human consciousness, a spirit of fertile interdependence and informated collaboration that will trickle down to the problematic Earth populations left behind. A compendium of recent work in space settlement planning, *Living and Working in Space* promotes the use of the behavioral sciences in mediating a technological environment for human habitation. As part of the space team, anthropologists, psychologists, and sociologists will maintain continual surveillance of human bodies, studying reproduction, sleep cycles, time sense, physical and mental stress, and the effects of weightlessness, isolation, and noise. "Artificial life" may produce time sense warps, "psychotic reactions," "spatial illusions," interpersonal conflict, depression, boredom, "anger displacement," a "need for dominance," motion sickness, water retention in the face, and a loss of body mass.[75] In addition, conflict among disciplines, cultures, and ethnic groups might arise.

The answer to controlling these human disturbances in techno-utopia is the application of James Grier Miller's general living systems theory, a complex symbol language of subsystems and processes. In a space environment, bodies become ingestors, distributors, converters, producers, extruders, and decoders, components in a biotechnical system for control of matter, energy, and information flows.[76] Thus, differences are transcended as humans become synergistic, ergonomically conditioned components in the metamachine. Here, the "informating" of knowledge workers in a postindustrial economy based on instantaneous communications, erosion of managerial hierarchies, the formation of strategic alliances and teams in electronic exchanges, and the potential for virtual universities and corporations is given stability under the rubric of mission success and safety.[77] Living systems theory provides the paradigm for a new, entirely planned microculture that will determine every facet of a spacefarer's existence, from decor to diet, from language to sex, for harmonious system functioning. For example, *Living and Working in Space*, sounding much like an L. L. Bean or Land's End catalog, extols space shuttle fashion: a "custom fitted, cobalt blue, soft cotton, lined zipper jacket and pants with coordinated blue shirts," having the functional attraction of being fireproof; other suits are of "light and heat reflecting metallic mylar which also serves to protect from meteorites."[78] Space planners stress the safety of their rationally managed synthetic biospheres that include "storm shelters" for protection against solar flares.[79] As longtime space consultant David Criswell explains in his discussion of future space biospheres, inhabitants of "s'homes" (space homes) will have "feelings of safety [that] reasonably spring from the certain knowledge that their advanced technologies constructed, operated, and can constantly refine their places and build new ones."[80] Space planners construct the s'home as a protective technological matrix where consensual decision making takes place among productive knowledge workers devoted to mission safety.

Biospherics

This coordinated, safe, and benevolent lifepod for the new cybernetic worker has antecedents in the public fallout shelter and other enclosed emergency environments planned and implemented in the early 1960s, just when the U.S. space program was well under way. The public fallout shelter system was designed to preserve urban workers for an imaginary postapocalyptic recovery from nuclear war. The U.S. Office of Civil Defense (OCD) policy urged labor leaders to contribute to the survival of the free enterprise system by registering their "emergency manpower," which might then be relocated to safe shelter. In its pamphlet *Labor's Role in State, County, and Local Civil Defense*, the OCD outlined the "personal responsibilities" of workers in fallout shelters: cooperation with authorities, self-discipline in the performance of assigned tasks, and emotional control.[81] Government pamphlets routinely pictured white business executives with briefcases entering public shelters, but used a paternalistic, sociological rhetoric aimed at the working class. Charles Fritz, a psychiatrist working for the government-sponsored Disaster Research Group, advised that shelter activity be channeled into "socially useful form": public shelters "should be combined with productive facilities (factories and work-shops), so that the period of shelter stay can be used in the active manufacture or production of goods."[82] To repress any pathological worry about the future, Fritz encouraged stocking shelters with hand tools and raw materials for "the production of useful survival items."[83] Herman Kahn recommended that all urban workers in the postapocalyptic environment be provided with radiation monitors; he wrote, "Assume . . . that a man gets sick from a cause other than radiation. Not believing this, his morale begins to drop. You look at his meter and say, 'You have received only ten roentgens, why are you vomiting? Pull yourself together and get to work.'"[84] Under the threat of annihilation in this hallucinatory scenario, workers would depend not only on instrumentation but on the experts who could read it. With a marked and monitored body, pathological worries could be dismissed, maintaining a smooth flow of pro-duction. In the redundant underground city, all human activity could be geared to the nation's preservation, under the slogan, "Survive, resist, recover, and win."

The fallout shelter program perpetuated and refined an institutional ide-ology that stressed control, surveillance, and punishment. At Michigan State University, the School of Police Administration and Public Safety provided the OCD with a "law and order" training manual for fallout shelter manag-ers. "Law and order" duties included protection of life and property, preven-tion of "unauthorized entry or exit," detention of prisoners, suppression of rumor, and rigorous repression of "troublemakers," "deviates," and others who lacked the proper "respect for order." Police officers were to be "the eyes and ears of the shelter manager and his staff."[85] Order and control, taken to absurd imaginary lengths, were particularly centered on suppression of de-sire, especially sexual desire, and thus control of reproduction. Civil defense consultant Robert Suggs, in his *Survival Handbook,* wrote that social values would crumble unless the shelter manager provided a "good example." In shelter tests, Suggs warned, "necking and overt sexual play were indulged in by teenagers; adults used obscene language in increasing amounts; gambling began."[86] One social hygienicist, Charles Clark, predicted that gonorrhea and syphilis would increase 100 percent after an A-bomb attack because of a "reckless psychological state," especially among young people. Clark suggest-ed massive stockpiling of penicillin and strict moral policing by clergymen.[87] Within shelters, physical contact was to be rigorously condemned. Public shelter promoters insisted that proper "management" would maintain order. As late as 1984, a "24-hour watch" was official fallout shelter procedure.[88]

Under these circumstances, the fallout shelter became a site of rigid surveillance, as the household was enclosed in a microenvironment and subjected to public management. Civil defense researchers carried out experiments involving prisoners, who were tested for psychological stamina under shelter conditions. One experimental shelter was surrounded by observation walks and contained two-way mirrors, television cameras, and audio equipment sensitive enough to pick up any conversation.[89] These experiments were tied to Navy-funded isolation studies of soldiers and researchers wintering in Antarctica and oceanographers living in underwater research installations, research projected to psychological and physiological conditions of humans in space capsules.[90] These initial studies emphasized the choice of appropriate space workers through various screening devices, workers who could stand the isolation of technologically enclosed environments and perform skilled physical tasks in transparent harmony with machines.

Some spaceologists went so far as to suggest cybernetic enhancements for space workers, such as a biologically implanted radio receivers and pumps for releasing psychoactive drugs and other medications, but these suggestions were generally received with disfavor in the popular scientific press, where it was argued that the cyborg was no longer human, and therefore would not represent a projection of humans into space. As prolific aerospace writer Martin Caidin puts it, "the solution of the cyborg as man's response to surviving the environment of space . . . is no more acceptable than to place entirely in the hands of computerized logic the decisions that affect our survival in a world of men."[91] On the stage of space travel, scenarios of the automated workplace were imaginatively enacted, establishing a reassuring national discourse on the survival of the laboring body under strict but benevolent observation, safely preserved but unpenetrated by machines. By the 1980s, this discourse had shifted to a reimagination of the informated body as a self-monitoring, physiologically and genetically intact conduit for abstract flows of energy and information.

The culmination of such isolation studies was the expensive, independent capitalist enterprise of Biosphere 2—part corporate research facility, part amusement park, and scaled-down lifeworld extraordinaire in which eight biospherians (all white professionals) survived for two years in a designer ecosystem. Despite its claims to have preserved nature under glass, Biosphere 2 is, as political scientist Timothy Luke points out, "essentially cybermechanistic," a "denatured space" replicating "the cyborg planet earth being constructed by transnational capitalism."[92] Animal, plant, and human inhabit-

Participants in the first mission to live inside Biosphere 2.
(Photo by Tim Fuller. Courtesy of Biosphere 2 Center, Inc.)

ants were chosen for their multitasking flexibility and ability to "handle the work load."[93] Advocated in a 1986 report by the National Commission on Space dedicated to the *Challenger* crew, Biosphere 2 was clearly intended as a prototype lifeboat for replicating and escaping a planet doomed by pollution and overpopulation. As founder John Allen, in his poem "Vertebrate X," lyricizes about the project,

> And we stream in and out
> of sunlit local gases
> on our sensors of delight
> at the same time collecting and recollecting Memory
> for that shipwrecked moment
> when we need it all and then some
> to keep the voyage going going
> and never gone.[94]

Before its exposure as a failed experiment lacking proper scientific controls, Biosphere 2's avid supporter, popular science writer Dorion Sagan, argued that artificial biospheres, or "baby earths," are the human future: "Without the shelter afforded by such enclaves, human beings, despite our pervasiveness on Earth . . . could become an endangered, a self-endangered species."[95] Imagining a world of these independent artificial biospheres dotting a planet made uninhabitable by pollution, Sagan cites biologist Garrett Hardin's tragedy of the commons, a theory that common resources are inevitably depleted through global competition. In the early 1970s, Hardin argued for a "lifeboat ethics," which he called a "special application of the logic of the commons."[96] Nations were like lifeboats afloat in a hostile world, and rich lifeboats could not afford to share space or resources with people in poor lifeboats. However, rich nations could impart knowledge of birth control and a sense of responsibility to these poor lifeboats to help them be more self-sustaining.

Biosphere 2 was a pure expression of this fantasy lifeboat for a postindustrial knowledge class: a containment of very scarce resources, supposedly without any exchange of physical goods outside the sheltered ecosystem, and closed to any species migration. However, the project's educational function in dispensing knowledge through technical links was raised to a nearly mystical status. Former space shuttle astronaut Joseph Allen describes his encounter with the biospherians as occurring through speaker phones, with a "rite of greeting" that involved matching hands with them against a "conversation window."[97] Knowledge of the self-referential system was to be communicated through windows, imparted to schoolchildren through video links, posted to listservs, and shared with scientists via telephone, all technologically mediated forms of communication behind which the body might be safely contained. Only solar energy and information necessary to species survival (internal and external) would be allowed to pass. Furthermore, like the fallout shelter, Biosphere 2 would serve as a cultural and biological archive. In this case, instead of the Bible and the Boy Scout handbook (standard inclusions in the fallout shelter library), it would contain a highly selective archive of phenotypes and technical information, integrated into the domestic milieu of the s'home and available for outside study. This highly managed leaking of public information between Biosphere 1 (Earth) and Biosphere 2, according to John Allen, would create a noosphere, which he defined as a "new world of intelligence" in a "massive exchange" of "words, numbers, sounds, images, procedures, patterns, at rates too quick and on a scale too vast to comprehend."[98] Even though physical bodies were continually touted as the best sensors for detecting potentially disastrous imbalances in the biosphere, these sensual relations were transcended by abstract, disembodied communications. Despite the enclosure of sensual bodies, Allen insisted that Biosphere 2 was somehow above being an isolation study because of data flows driven mostly by public relations. Yet isolation was the apparent key to creating a designer microenvironment that would carry knowledge workers, in matching wool jumpsuits, safely through the hostile environments of interplanetary space or the heavily polluted hothouse of Biosphere 1.

The fantasy of the biospherian lifeboat involved a protective containment of carefully rationed resources against external dangers, but depended on the continuing threat of internal disaster, mainly the buildup of carbon dioxide through the greenhouse effect. Furthermore, resources were purposely limited, exaggerating the dangers of scarcity. These potential disasters were intentionally constructed as integral to the artificial biosphere, demonstrating an internal as well as external mastery of them. Because the critique imparted through the *Challenger* and other catastrophes suggested the inevitable hazards of technosystems, Biosphere 2 set out to prove that such disasters as the greenhouse effect could be controlled through fail-safe technologies under constant human surveillance. As biospherians Abigail Alling and Mark Nelson explain in their account of the experience, disasters such as nuclear meltdowns and oil spills might be dismissed when they are at a distance, but "in a small living system, it is vividly clear that everything we do has immediate and potentially devastating effects on our surroundings . . . so we must police the environment."[99] Biosphere 2 was supposed to prove that famine could be prevented through careful rationing and the greenhouse effect could be managed through careful monitoring and stewardship by carefully screened, multitasking information workers.

Ironically, and inevitably, the system failed anyway because the designers did not anticipate that concrete structures would absorb and release greenhouse gases, or that one of the biospherians would slice off the end of her finger, or that cockroaches would proliferate—all unpredictable disasters that consistently undermine the most rational structures. Furthermore, Biosphere 2's managers cheated under the pressures of maintaining financial viability, and pumped oxygen into the facility, opening them to charges of corruption. In 1994, Biosphere 2's financial backers attempted to legitimate the project by shifting its control to Columbia University. *Scientific American* described the facility as an "ersatz Eden" with "wildly proliferating vegetation" searching for a "scientific identity."[100] Rational management might once again domesticate the disastrous Biospheric archive through cleanup, observation, and cataloguing by the proper external authorities.

Lifeboat dreams such as Biosphere 2 are not purely escapist fantasies, but reproductions of disastrous conditions under apparently controllable circumstances. They imaginatively contain the disaster within narrow, conceivable parameters. As an expression of survival, they simply reiterate and summarize the technocultural formations that produce disaster in the first place. Such microenvironments present a destitute, fatalistic, homogeneous vision of a cultural future, despite their promoters' soothing rhetoric of domestic safety and harmonious social order. Survival becomes containment, a reign of threat under which rationing, control over movement, forced production, sanctions against intimacy, and constant surveillance are normalized. Workers participate willingly for protection against a destruction that is produced by the structure itself. For despite the rhetoric of a great ship launching humans into a utopian future, the lifepod is a disaster waiting to happen.

2. Sarcophagus

Pyramid of Power

In 1989, the Department of Energy (DOE) initiated an interdisciplinary study to develop a marker for the politically besieged Waste Isolation Pilot Plant (WIPP), a salt cavern in New Mexico excavated in the 1970s to hold nuclear waste from government defense projects. Because of potential problems with the geologic formations surrounding the dig and new regulations for transuranic waste disposal, the plant has never been put into operation. The greatest concern for radioactive leaks into the surrounding environment involves "inadvertent, intermittent human intrusion"[1] through exploratory bore holes that might penetrate the hold and the pressurized brine underneath, forcing radioactive isotopes up into the atmosphere. Thus, the EPA requires markers or other warning systems that will give information about the site for 10,000 years into the future. (The time span is arbitrary and does not represent a limit to the site's danger because some transuranic elements have a half-life of at least 25,000 years.)

To fulfill the EPA standard, Sandia Labs (under the direction of the DOE) called together teams of knowledge workers to extrapolate possible future cultures and scenarios involving WIPP and to design the marker system. With backgrounds in a wide variety of disciplines, from sociology to materials science, many of them had worked on the Search for Extraterrestrial Intelligence (SETI), including Jon Lomberg, chief artist for Carl Sagan's *Cosmos* series, who helped design the *Voyager* plaque, and Ben Finney, an anthropologist who sailed a replica of an ancient canoe from Hawaii to Tahiti using traditional Polynesian navigational techniques and then applied his theories of human evolution and migration to space travel. The primary question for the teams was how to communicate potential disaster to future people who would surely no longer have our language, culture, technology, or even physical form. As Finney later put it, in a lecture predicting the inevitability of human spacefaring, "How can we—despite our newly found scientific knowledge—forecast all the evolutionary developments that will follow over the next five million years as our descendants disperse among the stars?"[2]

Despite the acknowledged impossibilities inherent in such forecasting, the marker design teams came up with ideas for menacing structures to cover the cavern with stone spikes, heat-absorbing black concrete surfaces, nearly impenetrable fields of boulders, and a square marked by a map of all nuclear waste sites. A monument to a disaster that had not yet occurred, the structure would serve to show a site "indelibly imprinted by human activity,"[3] that is, indelibly imprinted by culture, and an exceedingly dangerous and menacing one at that. With names like Black Hole and Landscape of Thorns, these designs were a review of national identity in the post–cold war era, when the government's massive technological projects, which had come to define that identity, were being subjected to harsh critique. Revelations of the DOE's terrible mishandling of nuclear and chemical wastes and secret experiments, including the infamous Green Run at Hanford, made nuclear weapon building particularly suspect, especially in communities surrounding DOE installations. The monumentalization of waste containment created a metanarrative that suppressed local critiques in a grand futuristic scheme to create an archive, preserving cold war nuclear images and texts as vital to future survival.

The "last cold war monument,"[4] as science journalist Alan Burdick calls it, was a symbol of the national border that both contained an internal production of danger and prevented alien invasion, for surely humans in 10,000 years would in many ways be "alien." During the first phase of the project, futurists met to predict possible societies that might penetrate the site, inventing a government-sanctioned science fiction of alternative futures as a basis for

Spike Field design for the WIPP Marker.
(Photo by Sandia Labs/Michael Brill and Safdar Abidi)

marker development. Their prospective societies reflect fractures in the Western project of technological development: contemporary anxieties over gender, race, and ethnic differences and challenges to scientific epistemologies. Assessing the chances of people stumbling onto the site, many of the scenarios are sober predictions based on a wide spectrum of change, such as increases and decreases in knowledge of the site, nuclear expertise, population density, and resource consumption. But some are more elaborate fictions that describe possible misinterpretations of a marker message system. One of the stranger scenarios involves a world taken over by feminists, in which "twentieth-century science [is] discredited as misguided male aggressive epistemological arrogance."[5] Thus, the site warnings are dismissed, and the feminists begin mining in the area. Other futures include the rise of religious cults hostile to science, a loss of knowledge about the site because of war and rebellion (including the secession of New Mexico), the looting of markers, the rise of an oral or visual culture with a subsequent loss of literacy, and an invasion by virus-ridden robot workers who ignore the marker system.

To save potentially misguided inheritors of the cold war's radioactive landscapes, the marker designers sought symbols that could supposedly still be read in 10,000 years. One of the teams borrowed its iconography from *The Scream* and an expression for nausea considered universal by some sociobiologists. The lasting meaning of Mr. Yuk was debated, as was the skull and crossbones. The icons represented a first level of meaning that would be accompanied by three other increasingly complex layers of information about the site, the fourth intended for engineers and scientists, "as well as historians and archeologists who wish to study 20th–21st century culture."[6] This archive would contain information about symptoms of radiation sickness, the site's geology and contents, the half-lives of the buried radionuclides, maps of all nuclear repositories in the world, star charts, and rationales behind the marker system, including a discussion of its redundancy and choice of icons and languages.

The discourse surrounding the marker was full of fears about the perpetuation and reproduction of culture, besieged by leaking and penetrable borders, temporal decay of structures, and knowledge not passed down to future generations. This was to be an anxious, uneasy monument in which culture was built around an always potential disaster. But perhaps, as Burdick ruminates, the marker would come to signify only itself, a "cultural landmark" to the power of the United States, like "Lenin statuary," and thus subject to a revolutionary toppling.[7] What if the marker, as Jon Lomberg suggests, no longer referred to any danger, but was perceived as art that would draw people to it? Without the logic of the nuclear state, this cultural project, intended by some team members to be "a gift from our century to the future" or "environmental sculpture,"[8] would not mean anything at all.

At the same time that the U.S. post–cold war imaginary was being projected onto the WIPP site, another nuclear monument was beginning to show cracks. During the Chernobyl[9] Power Station's Reactor 4 containment, at great risk to their lives and often unable to see what they were doing, hundreds of workers, wearing lead plates and operating remote-controlled equipment, hastily constructed a concrete and steel encasement to shore up the building's remains, enclosing tons of radioactive waste. The structure was equipped with various sensors to monitor potentially dangerous fluctuations in vibration, temperature, humidity, and radiation. This encasement, called simply *Ukryttja*, or Shelter, in Ukrainian, became widely known as the Sarcophagus.[10] Some observers extended the metaphor to discussions of the valuable "corpse," the melted nuclear core that lay like a "modern-day pharaoh, a pharaoh still alive,"[11] ensconced in an expensive pyramid that must remain intact for many thousands of years.

Thus, the "voiding-induced super-prompt critical power excursion"[12] that sent tons of radioactive particles into the atmosphere had a visible monument, one that government officials, nuclear scientists, and sympathetic journalists could critically construe within conventional nuclear images of veiled secrets, hidden wealth, barely harnessed forces, and monolithic authority. In 1970, in his critique of the "megamachine," Lewis Mumford argued that those nuclear reactors, in particular, were the "new pyramids" of a technocratic priesthood.[13] From the early days of atomic energy research and development, nuclear physicists had used a discourse of transformation, rebirth, and spiritual force to sell their nuclear projects. By the 1980s, this rhetoric had become so firmly implanted in popular media and political speeches that some scholars viewed nuclear energy as a mythic power, a numinous "Dr. Strangegod"[14] that loomed over the world, inciting awe, blind worship, and "missile envy."[15] In 1988, relying on Jungian theories of collective archetypes, science historian Spencer Weart wrote that nuclear energy had become a deep psychological symbol of a "transcendental power" and a "new Arcanum, permanently connected with the most terrible, fascinating, and sacred of all human themes."[16] This nuclear rhetoric shared the conventions of the technological sublime, used in modern national narratives to excite public approval for certain trajectories of technological development.[17] However, rather than some deep-seated psychological manifestation of a collective unconscious, nuclear images are persuasive deployments that rely on familiar rhetorical devices. As massive concentrations of capital, labor, materials, and equipment, cold war nuclear weapons and nuclear energy facilities required continual solicitations of local and national support, including emotional appeals that made the technology seem autonomous, essential, omnipotent, and unassailable, evoking awe and terror. As the tomb of the nuclear pharaoh, the Chernobyl Sarcophagus held this legacy, but became a leaky monument, unable to hold its containment borders, either physical or symbolic. Political and environmental critiques, stemming from a well-established antinuclear protest movement in the West, undermined any gestures toward a cultural revival based on nuclear systems building.

Whereas nuclear weapons maintained their sublime aura in the late cold war climate, even among antinuclear protesters, the Three Mile Island near-meltdown in 1979 seriously damaged the credibility of domestic nuclear energy producers, nearly ended nuclear power plant construction in the United States, and threatened it elsewhere. The Chernobyl meltdown had even broader implications for nuclear states. Not only was it a much bigger disaster, involving a truly catastrophic system failure, explosion, and massive release of radio-active materials and toxic chemicals, but it had sweeping, unprecedented environmental and sociopolitical effects. For one thing, the Chernobyl disaster provided opportunities for an effective vocal dissent against an increasingly destabilized Soviet government, with broad repercussions for the West, including the United States. Some have attributed the fall of the Soviet government, in part, to the mobilization of antinuclear organizations.[18] And certainly Chernobyl created conditions for opportunistic dissent. Adriana Petryna writes that among Ukrainians, Chernobyl became "a consuming hole of the present, a rupture in historic time, systems of belief, and representation."[19] The Ukraine Energy Minister at the time of the accident, Vitali Skliarov, later cajoled the readers of his "shocking firsthand account": "After Chernobyl literally everything—our thoughts, perceptions and views—have to be fashioned anew."[20]

Although many witnesses to disaster often describe this experience of a temporal fission when their worlds changed forever, the Chernobyl effects expanded beyond the local into an international media spectacle that derived its apocalyptic tone from a heritage of nuclear discourse. To give some sense of scale, agency officials and journalists often compared the Chernobyl disaster to Hiroshima, the apocalyptic event of the twentieth century. Radio-active emissions were compared to those released from the atomic bomb,[21] and Japanese scientists who had studied the *hibakusha* were engaged for Chernobyl research. But more socially powerful were the familiar nuclear representations, such as concentric circles radiating from Ground Zero, a postapocalyptic wasteland with relocated populations, blackened bodies with burned and tattered skin, and military forces mobilized to contain an invisible radiation that seeped through household cracks and into cow's milk.

Although the threat of radioactive exposure was real enough as the Chernobyl cloud filtered unpredictably across the northern hemisphere, its contamination consisted not only of transuranic particles, but also of information itself, information about toxic radiation effects, aging refugees escaping technological catastrophe, strontium-tainted milk as far away as Italy, radio-active reindeer buried in pits, whole tracts of pines burned red, political turmoil, and deserted cities. Detected by a Swedish nuclear power plant and U.S. satellite imaging, Chernobyl demonstrated the power of new communication networks to facilitate uncensored information flows[22] and the failures of modernist technological systems. Escaping the Soviets' attempts to censor information, Chernobyl's revelations all involved porous borders: the reactor spewing hazardous materials beyond its walls, a radioactive cloud that crossed into other nations, radioactive particles that penetrated bodies and altered genetic codes, data flowing through lines of secrecy, and crumbling infrastructures.

In the wake of this fissioning of established orders based on faith in technological safety, national boundaries, protected information, and contained systems, the disaster suggested a catastrophic limit to technological development at a time when "you can't stop progress" was a mantra for postindustrial states. Eleven days after foreign journalists were allowed into Kiev to investigate the accident, *U.S. News & World Report* announced a new "high tech anxiety," comparing Chernobyl to the release of methyl isocyanate at Bhopal, the *Challenger* space shuttle disaster, and the explosion of a *Titan* missile, and identifying other dangers in aviation control, chemical storage, genetic engineering, landfills, building foundations, computer security, and electrical plants. Warning that in three years the number of Americans who "still say that science and technology do more good than harm"[23] had fallen from 83 to 72 percent, the article generated the very anxiety it claimed to identify.

The sweeping scope of technology associated in this way with Chernobyl made it an icon of failed sociotechnical systems, including everyday interactions within those systems. The August 1986 meeting of the International Atomic Energy Agency, devoted to Chernobyl, adopted recommendations for their future activities that included more analysis of "the balance between automation and direct intervention by an operator" and the "man-machine interface."[24] Furthermore, descriptions of the reactor design, written for a lay audience, domesticated the dangers. For example, science fiction writer Frederik Pohl, in his novel *Chernobyl,* describes the accident as "no core meltdown," but a "simple matter of carbon combusting in the presence of oxygen, not basically different from the blazing logs in the fireplace of a split-level ranch house."[25] And David Marples, who has written several sensitive studies of Chernobyl's effects on Ukraine and Byelorussia, compares the reactor to "a simple gasoline-powered portable generator that people use in cottages and on trips."[26] A typical risk discourse on hazards that compared the dangers of radiation exposure to driving automobiles and smoking also served to map Chernobyl into familiar settings. Even Richard Rhodes, an eminent nuclear historian, wrote a defense of nuclear power that placed Chernobyl radiation dangers in the context of risk assessment, using nuclear advocate Bernard Cohen's list of supposedly comparable hazards such as radon, fire, coffee, firearms, peanut butter, airline crashes, occupational accidents, and birth control pills.[27] Like the *Challenger* disaster, Chernobyl was variously read into ordinary contexts of work and domestic life. This familiarizing of an extraordinary event, under the guise of technical education for naive, fright-ened people, was double-edged. On one hand, it tamed the disaster into com-prehensible, manageable terms; on the other, it made ordinary life seem a place of catastrophic potential, with nuclear fireplaces and menacing peanut but-ter. As the cold war threat of nuclear holocaust had already made the civilian workplace and the home potential loci for apocalyptic technological violence, Chernobyl made the dangers of unpredictable fires, explosions, and radioac-tive contamination seem intimate, ordinary, and inescapable, part of every-day bargains with technosystems.

Thus, the sociopolitical stakes in containing the disaster were high, not only for the nuclear industry, but also for nuclear states with heavy investments in complex technologies. In 1989, in his history of modern American inven-tion, Thomas Hughes wrote that the twentieth century has been driven by a technological momentum, fueled by nuclear weapon development and re-sulting in large, rigidly managed systems of production. He predicted that the abandonment of large systems and the subsequent arrival of a "postmod-ern era" would result from "a confluence of contingency, catastrophe, and conversion."[28] Chernobyl represented that very confluence, revealing the shifts and gaps in such systems. Because it came on the heels of the *Challenger* disaster, when joint U.S. military and corporate technological developments were already subjected to intense debate, the Chernobyl accident reconfirmed the dangers of such systems. U.S. nuclear energy and weapon projects came under a renewed scrutiny, especially as revelations of the DOE's secret radi-ation experiments, inept plant operations, and mishandling of nuclear waste were coming to light. In the press, there was an immediate, opportunistic, and short-lived political effort to distance the United States from the Cher-nobyl meltdown. The attempted establishment of an official U.S. point of view is well represented in a *Time* magazine cover appearing some two weeks after the accident. The cover presented the first official Soviet photograph of the damaged reactor, originally in grainy black and white, but doctored to glow in hot reds, oranges, and yellows. Underneath, the word "MELTDOWN" appeared in large black letters. In the upper right corner, framed in a red tri-angle and symbolically protected from any emanations from the reactor, was President Reagan, on a state trip to Bali, covered in flowers, appearing to be on a holiday, and laughing down, seemingly at the ruined reactor. His point of view, appropriately from the right, was a political wedge into the specta-cle, implying celebration, even though Reagan had sent his condolences and offers of aid to Moscow, along with a stern critique of the Soviet government's secrecy about the disaster and its consequences. But this political distancing could not be long sustained, and the next cover of *Time* put a Chernobyl radiation worker in Reagan's place, looking down on a birthday cake for Baby Boomers turning forty, surrounded by hula hoops and phonograph records, the domestic nostalgia of the cold war generation.

As part of their own damage control, the Soviets constructed the Sarcophagus as a virtual monument to sustained production; the other onsite reactors remained in operation, supposedly protected from any further damage. Alexander Sich, who, as an MIT graduate student, worked with Ukrainian scientists in the postdisaster investigation of Reactor 4, remembers that the Soviet government promoted the Sarcophagus as "the most visible and attention-drawing symbol of triumph over the accident."[29] Ukrainian-American ethnographer Adriana Petryna suggests that the Sarcophagus evoked Lenin's tomb and was meant "to incite a sense of physical, moral, and spiritual rejuvenation within the Soviet population."[30] (The Chernobyl nuclear power station was originally named after Lenin.) Following its construction, Soviet authorities heralded the Sarcophagus as a testament to engineering and heroic plant operators and firefighters. A plaque on the building's outside wall pays tribute to these Chernobyl martyrs, who willingly went down with the reactor, like the boiler operators on the *Titanic*. Officials overseeing the Reactor 4 containment often described their work as a highly successful, highly complicated military exercise, involving a massive concentration of soldiers, medical personnel, building supplies, and equipment, including robots and remote-controlled bulldozers. The epicenter of the disaster—a physical and symbolic hole in the middle of nuclear state projects—was seemingly replaced by a technological marvel.

The Sarcophagus not only reinstated a military hegemony and state pride, but proved a useful monument to science. The structure was supposed to last for hundreds of years, as its complex data-collecting instruments continually monitored any fluctuations of temperature and radiation.[31] Through surveillance possible only through instruments, scientists and technicians were supposed to contain and supervise the effects of reactor meltdowns. This, at least, was the rhetoric of the Soviet nuclear establishment: one official argued that "science requires victims."[32] Despite reassurances that the nuclear reaction had been bound and the "China Syndrome" prevented, authorities had very little knowledge of what was actually going on inside their pyramid. In 1987, having no idea of where the migrating fuel (including 641 kilograms of valuable plutonium) might be, a scientific team began penetrating the reactor's basement chambers with boreholes and teleoperated, video-equipped robots. According to journalist Piers Paul Read (best known for his account of cannibalistic plane crash survivors trapped in the Andes), the Chernobyl Complex Expedition, called the "Stalkers," found lava flows with "strange new crystalline forms," masses that turned out to be fuel trapped in liquefied, hardened silicate, like "a fly in amber."[33] This "corium" contained 71 percent of the reactor's fuel, along with stainless steel, serpentine, graphite, and other substances from the reactor explosion, now transformed into glass and ceramic stalactites, stalagmites, and hardened streams.[34] One of these flows was named the "elephant's foot," suggesting the taming of the beast in a concrete zoo. In narratives of the team's site penetration and discovery of the "Chernobyl lavas," the interior of the Sarcophagus was constructed as a place of wonder and fearful mystery, an awesome abyss, accessible only through virtual means. One of the first two scientists to enter the reactor's ruins describes his continuing passion for probing the Sarcophagus and his experience of the Chernobyl lavas: "It is very black, like glittering coal. . . . It's awfully beautiful, it shines like silver. You feel like you're on the moon but rationally you know the radiation is extremely high. It is an unforgettable experience."[35] Other scientists see it as a means for proving their manhood, or find the thrill of exposure to the lavas similar to a drug addiction.[36] Western reporters describe visiting the Sarcophagus as an Indiana Jones adventure to the nuclear lava: "In places, the lava wells up eerily from the floor. One such 'stalagmite' looks like an elephant's foot. Curiosity is dangerous, though. Every second you linger, you're getting zapped by radiation."[37]

The cultural resocialization of nuclear energy transformed the containment site into a science museum of alluringly grotesque forms, holding the secrets of the Chernobyl meltdown and thus the historical mistakes of the old state. However, as Sich points out in his critique of the Soviet and later Ukrainian establishment, meaningful scientific research was and is constantly hindered by economic and political interests.[38] Furthermore, the structure has proven to be ineffective. Reports from the scientific team working in the containment structure reveal that the Sarcophagus's walls are cracking, allowing rainwater to seep in, posing the possibility that radioactive particles are leaching out of the ruined reactor.[39] Rapidly mutating voles, mice, birds, and insects are nesting in the building, and may carry radioactive particles out. The upper biological shield, named "Elena,"[40] is hanging precariously and may collapse and severely damage the building. (Fuel elements thrown several meters from the reactor are called "Elena's hair," evoking some conventional associations between female sexuality and atomic energy.) Several hypothetical scenarios involving earthquakes, tornadoes, and plane crashes suggest that the Sarcophagus might be breached in another catastrophe. And no one really knows whether particular concentrations of remaining fuel (currently subcritical) and other elements might start a chain reaction or cause an explosion, spewing even more radionuclides into the atmosphere. As one Sarcophagus overseer explains, "Everything that is inside has not been fully studied or understood. We do not have a sufficiently complete control system."[41] Plans are now under way to build a very expensive "Shelter-2" to envelop the old Sarcophagus or transform it into a concrete-filled monolith. At a conference to raise money for the reconstruction, held in New York in 1997, Vice President Al Gore called the project a "new historic journey for a more secure and safer future for Chernobyl."[42]

The current Sarcophagus is mostly symbolic, a shaky virtual containment promoting a technocratic worldview to hundreds of Western tourists who flock to the scene to be thrilled at the radiation risk and to be reassured that the technicians have civilization's dangers under control.[43] It is now common for technological disaster sites to become tourist attractions in which local guides are employed to soothe visitors with narratives of security and safety. For example, well over a half a million people have toured the Three Mile Island tourist center to learn of reactor safety,[44] and in Valdez, Alaska, visitors can take a bus tour of the pipeline terminal from which the *Exxon Valdez* sailed, as the driver points out its safety features. Site workers and tour guides, who mediate this experience for visitors, also demonstrate the new terms of living under controlled technological threat. Tourists can visit the Chernobyl site, talk to workers there, and receive a calendar with the slogan, "Safety culture, effectiveness, social progress."[45] A *Newsday* journalist reports that one Chernobyl plant tour guide, after being asked whether he feared radiation, replied that he did not even look at his dosimeter: "I don't want to know what it reads. . . . If I reach my yearly exposure, they wouldn't let me work here. So if I'm here, I must be OK."[46] Noting the familiar tradeoff between economic expediencies and dangerous working conditions, a plant foreman told an Associated Press reporter, "There's a little risk working here. . . . But how could I not work here for that kind of money?"[47] The deputy manager of the Sarcophagus, who leads tours into its twisted wreckage, often preaches optimism and dismisses any fears of health effects from the radiation, which he calls "the best in the world."[48] And like the temporary inhabitants of Biosphere 2, the stalkers rely on sensory evidence of danger, leaving behind their radiation monitors and relying on taste to detect radiation. If their monitors read too much, they might lose their jobs.[49]

The Sarcophagus worker's positive attitude in the face of certain physiological danger and his willingness to brave it echo with a broader official discourse on radiophobia. As Marples reports, in the early days of the disaster, Soviet health officials suggested that people were suffering from a psychological disorder, that fear of radiation was consuming people's minds and making them sick, and that these fears were contagious.[50] Therefore, the victims of Chernobyl were blamed for their own condition. Although official agencies have subsequently distanced themselves from the original use of *radiophobia* as a derisive term, most still contend that stress from relocation, radiation fears, or the larger breakdown of Soviet society, has caused much of the increase in illness in Chernobyl evacuees. The World Health Organization (WHO) suggested in 1995 that a whole host of ailments, such as diseases of the nervous, digestive, genitourinary, and endocrine systems, might be attributed to stress from radiation fear, enhanced by "lack of information immediately after the accident."[51] A year later, a joint conference of WHO, the International Atomic Energy Agency, and the European Commission reconfirmed that stress over the accident is causing anxiety, depression, and psychosomatic disorders. In their summary report, the agencies state, "It is understandable that people who were not told the truth for several years after the accident continue to be sceptical of official statements and to believe that illnesses of all kinds that now seem more prevalent must be due to radiation. The distress caused by this misperception of radiation risks is extremely harmful to people."[52] Furthermore, the report asserts that "protracted debate" over radiation health effects is making people sick. Thus, any kind of critique or even curiosity about the scientific uncertainties surrounding long-term exposure to radiation is made into a dangerous, self-defeating activity rather than a potentially empowering one.

In this and other self-interested institutional disaster study and prevention efforts, the disaster is seen as a problem of information and its dispersal. The disaster and its effects unfold in a virtual world in which information causes or prevents harm, as people read that information into their own bodies and actions. Radiation becomes an imaginary agent that is harmful mostly in its coding as information. Seeing clear lines of cause and effect between such information and "psychosomatic" bodily damage, institutions set out to educate populations that in reality may have their own agendas for the discovery, creation, and deployment of such information. This was never so apparent as it was in the case of Chernobyl, where anemic, guarded public health efforts to inform a population perceived as passive, if not neurotic, stood alongside the political mobilization of radiophobia as a significant means for social reformation. Ukrainian poet Lyubov Sirota, who was evacuated from the site, writes of radiophobia,

> It is—
> when those who've gone through the Chernobyl drama
> refuse to submit
> to the truth meted out by government ministers.[53]

Jane Dawson explains that after Chernobyl, in Russia, Lithuania, and Ukraine, "anti-nuclear publicists portrayed nuclear power stations as horrifying and real threats to people's health and the continued existence of a national, ethnic, or territorially defined group of people. Plans for nuclear power stations were equated with policies of genocide."[54] By 1991, these environmental efforts had diminished in the Ukraine as the economy declined and Ukrainians came to see nuclear power as a means for continued energy self-reliance.[55] Within such a milieu, any public health efforts to educate are politically charged, even if information is presented in the language of objective science. Furthermore, medical researchers are nowhere near to understanding the complex health effects of the accident—the immune system breakdowns now commonly known as Chernobyl AIDS. The bodies of Chernobyl's survivors have become highly politicized, bearing the contested symptoms of a complex technological and cultural assemblage in which cause and effect cannot be identified clearly. Nor can cultural influences be separated from the physical realities of bodies clearly marked by disaster.

In his role as mediator of the new state's technological systems, as well as curator of the old, the Sarcophagus worker asserts that one can cheat disaster and strike an attractive psychological bargain with it. In this world, nuclear technicians and scientists manipulate information and rationalize danger in their service to an irrational monument. Because in the postdisaster milieu no one can have blind faith in the rational, safe functioning of systems, a working survival within them depends on complicated instruments, both physical and psychological. However, although the Sarcophagus and its sister, the WIPP project, have an official capacity as institutional icons of protection and safety, they are layered with cultural meanings that are so contradictory that their messages all have a certain inherent irony. These monuments to safety are built around a continuing disaster that will unfold for thousands and thousands of years, barring some miraculous technological fix. And they stand in demarcated territories now considered experimental zones, where any assertions of certainty are undermined by incomplete understandings of polluted ecological systems and unfolding social changes.

The Zone

Immediately after the Chernobyl disaster, government officials established a 10-kilometer exclusion zone around the reactor, and then, three days later, a 30-kilometer mandatory relocation zone, now known variously as the Dead Zone, the Estrangement Zone, the Alienation Zone, or the Forbidden Zone. The 10-kilometer zone and the 30-kilometer zone are usually represented as two concentric circles drawn around the plant and the accompanying town of Pripyat. Within the following year, a government health commission had established four zones for evacuation and monitoring, but this was not made public until 1989. The most often reproduced visual image of Chernobyl's effects is still the two initial evacuation zones, despite the spread of radioactive particles throughout the Soviet Union and Europe, and perhaps as far away as Antarctica, with heavily contaminated areas at least 200 kilometers away from the site.[56] The designation of these zones was, in itself, an expedient containment strategy that belied Chernobyl's full global impact. As perfect circles, the zone demarcations spoke of a rational state order laid onto a rapidly transforming terrain, a geometry that belied the spatial complexities of unpredictable winds and uneven fallout distribution. The "consuming hole" was ineffectively heaped with sand, boron, and lead and surrounded by conventional images of radial targets and atomic bomb blasts represented as concentric circles in civil defense maps. But these arbitrary rings of defense haven't prevented radiation from leaking through them.

Two kilometers from the reactor, in the exclusion zone, stands Pripyat, the most well-known nuclear city after Hiroshima and Nagasaki. Over the period from the disaster in 1986 to its ten-year anniversary in 1996, a genre of Chernobyl writing has evolved that is at once a horror story, anthropological study, and adventure narrative into the apocalyptic zone with its nuclear tomb and "dead" town (actually occupied by Chernobyl workers). It is common for journalists, filmmakers, visiting scientists, and tourists to describe their brave journeys into the deserted, modernist, radioactive Pripyat, with its empty high-rises and cracked sidewalks. They send postcards from plague-town with its "gigantic rats and cockroaches."[57] Two months after the disaster, *Time* magazine writers portrayed Pripyat's "lifeless silence": "The only movement that suggests human habitation is the flutter of laundry on the clotheslines."[58] A year later, *Newsweek* called Pripyat a "ghost town": "A store is filled with dresses and canned goods, but no shoppers."[59] The town's rusting Ferris wheel, turning "riderless in the wind,"[60] is often mentioned, speaking of abandoned familiar pleasures.

In Nova's *Back to Chernobyl* (1989),[61] journalist Bill Curtis journeys into the zones as his companion, Harvard physicist Richard Wilson, checks dosimeter readings to assess their slowly increasing level of exposure. Suggesting that Chernobyl might be the "Hiroshima of our time," Curtis and his team travel down the snowy road in the "bitter Russian cold." Shots of Pripyat include abandoned prams, family photographs, clocks, paintbrushes, dishes and cups on a table, and a day calendar open to April 26. At the end of the program, Curtis stands against a backdrop of empty high-rises and asks, "Is

Deserted secondary school near Chernobyl. Illinsty, Ukraine.
(Photo by Greenpeace/Shirley. Courtesy of Greenpeace International)

this the future?" During her tenth-anniversary visit to the exclusion zone, a CNN correspondent found it "silent, its half-emptied houses showing evidence of a panicked flight, lying in memorial to the world's worst nuclear disaster."[62] Another anniversary visitor describes Pripyat's abandoned amusement park and "rubble-strewn apartments" with "skeletons of cats and dogs that were locked inside, awaiting their owners' return."[63]

The theme of abandoned artifacts, holding only an aura of living humans who had left them in mid-use, has been repeated time and time again, until hardly any Chernobyl travel narrative is complete without it. The narrative has hardened into a very familiar set of conventions, conventions that hold the disaster in a solitary moment, forever set with the ruined reactor's clock at 1:23:46 A.M. The clock reminded Glenn Alan Cheney, who traveled into the Forbidden Zone in 1991, of "the tower clock at Hiroshima that stopped at 8:15 on the morning of August 6, 1945."[64] Furthermore, frequent citations of Revelations 8:10–11, in which an angel sends the star of Wormwood to rain death upon the waters, have made the disaster a fulfillment of biblical prophecy. "Chernobyl" is named for wormwood, a medicinal plant found in the forests there. Many writers who seize the literary opportunity for evoking the supernatural and framing the event in traditional eschatological structures have taken up this coincidence. Descriptions of plants and trees, pushing up through the cracked sidewalks and concrete structures of the city, evoke a posthistorical world in which "nature can take its revenge."[65] As a former nuclear inspector on a tour of the city puts it, "Standing amid the broken glass and crumbling concrete of the Pripyat city hall, I looked at a Soviet mural portraying all the benefits of modern civilization—science, medicine and technology. How very hollow the promised benefits looked now, in the silent rows of cloned apartment blocks and the pavements where new birch trees were the only signs of life in that cemetery of a city."[66]

Despite the apocalyptic patina on Pripyat's remains, the repetition of these conventions proves that the dosimeter-armed, strictly monitored narrator can travel into the dead zone, observe the familiar wasteland contained in its frozen moment, and survive to retell the tale. Often, while evoking the horrors of the postapocalyptic ghost town, the traveler mentions his or her own vulnerability, the measures taken to prevent radiation overexposure, the donning of breathing filters and surgical caps, the increasing levels of radiation. Sometimes, this is presented in an oddly neutral tone: "A Geiger counter shows that near the plant one is exposed to a radiation level a few hundred times normal background."[67] Sometimes, the fear is more personal and imminent: "All you hear is the sound of your breathing, amplified by the nose mask. And that of the Geiger counter going click-click-click. And that of your heart, thumping in frantic syncopation."[68] And perhaps it is familiar: "The Geiger counter in my hand buzzes as if it is heavy television static."[69]

These Chernobyl travelogues are also survival narratives and, like the Sarcophagus tour guides, present strategies for coping in a world where death by technological means is more probable. The most famous of these trips was undertaken by physician Robert Gale, who became a minor celebrity when he went to Moscow with an offer to provide bone-marrow transplants to radiation victims. Later dramatized in a television movie starring Jon Voight and Jason Robards, Gale's book, *Final Warning,* is both an argument against nuclear weapons and an account of his journeys first to Moscow, and then to Chernobyl itself. Gale remembers that he first heard the news on the radio while he was shaving, after doing his morning exercises, while his children were sleeping. His evocation of an ordinary life, with its ordinary details, suddenly interrupted by a profoundly life-changing event, is very much in keeping with the conventions of disaster witness accounts. In this way, Gale immediately aligns himself with the victims of Chernobyl. However, the narrative makes frequent reference to Gale's running, running through Southern California, running "eight or nine miles at a seven-minute-per-mile pace"[70] through Moscow with a USA insignia on his clothing. The film version opens with Jon Voight, playing Gale, running down the street, listening to the news of Chernobyl on his radio. Thus, Gale's image is established as healthy, strong, physically fit, and American, in contrast to the weak, dying Chernobyl workers he met in the Moscow hospital. Another doctor, Orlov, who participated in the triage of Chernobyl firefighters, especially affected Gale: "Black herpes simplex blisters scarred his face and his gums were raw with a white lacy look like Queen Anne's lace caused by candida infection. Then, over several days, the skin peeled away and his gums turned fire-engine red like raw beef."[71] While suggesting sympathetic identification, the mirroring of Gale with Orlov also sets Gale's difference apart: his unaffected healthiness, his separation from the events he observed, his presence as an autonomous subject and narrator of his own life. In this way, Gale mediates the event for his American readers, demonstrating how they might perceive the catastrophe from a position of relative safety and physical integrity.

Furthermore, Gale's image is of a technological savior funded by capitalism,[72] a doctor who brought new, innovative life-saving devices to disaster victims. In a photograph in *Omni* magazine, Gale stands on a runway in front of a helicopter, hands on hips, stethoscope in his pocket, and earphones around his neck, in a white jumpsuit emblazoned with a caduceus and "MD Triage." The accompanying interview, conducted by cold war spy novelist Martin Cruz Smith, emphasizes Gale's use of "popular new technologies," mostly involving bone marrow transplants and cell cloning.[73] Gale performed marrow transplants on seven of Chernobyl's radiation victims, a procedure that later appeared to be of questionable value.[74] But Gale's image spoke not only of surviving the disaster, but also of making it *productive*. Thus, while Gale predicted the likelihood of a serious nuclear accident with mathematical certainty, he also promoted new technologies that might arise from the radioactive ashes of Chernobyl.

Gale's work was popularly presented as a worthwhile experiment, for the Chernobyl accident "created a kind of medical classroom—a unique if horrific opportunity to learn how to cope with large-scale exposure to deadly radiation."[75] Experimental medical technologies involving extraction, genetic manipulation, and replacement could replace destructive nuclear technologies as shared cultural projects. These new interpenetrations and modifications of the body, unlike accidental radiation damage, might be managed, controlled, and directed. In his *Omni* interview, promoting one U.S. business that donated thyroid testing kits to the Soviets, Gale noted, "It's ironic that we use radioimmunoassay with radioisotopes to detect exposure to radiation."[76] But that irony turns on an apparent contrast between a controlled, healing radiation and cancer-inducing radiation, one technology as a remedy for the other. Later, in the medical journal *Lancet*, Gale describes his return to Pripyat in the familiar terms of the Chernobyl tourist: "[It] is a ghost town of high-rise concrete buildings. Some public areas that I visited were particularly eerie—a kindergarten where about 30 cots were made-up, awaiting a mid-day nap that never happened."[77] But the lesson at the end of this bleak article is about technological conversion: He describes lasers developed for the Strategic Defense Initiative being used to detect genetic mutations in Chernobyl radiation victims. Declaring that technologies are neutral, Gale optimistically promotes conversion and cooperation in the use and development of nuclear energy sources.

In early Chernobyl discourse, the Western scientist, replacing the spy as the agent of discovery, played the role of a mediator who could convert the disaster from an annihilating technological force to a stimulator of technological progress. But these early suggested social cures paled in comparison to later ecological studies of the zones that have replaced apocalyptic destruction with revelation and a principle of plenitude, a flow of abundant (if somewhat frightening) forms from a radioactive source.[78] Four years after the disaster, geographer Peter Gould wrote that Chernobyl radioactivity had "served as a tracer, moving through physical and living worlds to disclose their chains of connection."[79] A UNESCO report, unusual in its emphasis on the interrelations of economic, political, psychological, physiological, and ecological changes, theorizes that the Chernobyl disaster served as a "stochastic causality" from which a new self-organizing system has arisen.[80] Radiation and other pollutants thrown from the reactor have created a permanently altered hot zone, "a vast man-made radioecological area within the biosphere,"[81] reforming geographic borders for relocated humans and animals migrating into vacated areas. Populations of living organisms have shifted as radioresistant and rapidly reproducing types such as voles and mice thrive while others, such as pines, weaken and decline.[82] With the evacuation of well over a hundred thousand humans, the zone is now being repopulated with red fox, gray wolf, moose, roe deer, wild boar, and feral dogs. In the transforming habitat, the absence of human impact currently outweighs radiation health effects. And the 10-kilometer zone has now been set aside as a wildlife preserve, devoted to radioecological study.[83]

Science illustrator Cornelia Hesse-Honegger's meticulous watercolors of "damaged" insects around Chernobyl and other nuclear accident sites render these zones as places of beauty in distortion. One of the most interesting of all Chernobyl representations, they present a baseline norm and its variations in highly detailed renderings: Symmetrical, "normal" bugs are placed next to asymmetrical mutations. Hesse-Honegger explains that she embarked on her illustration of bugs because she "wanted to find out how nature should really look."[84] According to the natural world constructed in Hesse-Honegger's work, nature is balanced and symmetrical before catastrophic radioactive incursions, and skewed and asymmetrical after. The controversy surrounding her exhibits lies in these assertions of the "normal" and her identification of an invisible, deforming agent: radiation. Although Hesse-Honegger claims a political side to her work, motivated by a desire to expose "humankind's senseless deformation of nature,"[85] the paintings aestheticize these mutations in a naturalist tradition in a way that photographs would not. Reviewers have called Hesse-Honegger's work seductively, disturbingly beautiful, as well as crisp, cool, and technically perfect.[86] A gallery press release for her After Chernobyl exhibition explains, "Her detailed studies are in the tradition of the stunning images produced by explorers, scientists and artists who recorded the wildlife of both chartered and unchartered regions around the world before the use of photography."[87]

Scientific descriptions of this strange new wilderness have been taken up by the media taking opportunity of the ten-year anniversary, and promoted as a transmutation of the disaster into an evolutionary force. Often featured are biologists such as Robert Baker, who studies small mammals in the zone and has found them to undergo a highly accelerated rate of mutation. Baker's descriptions of "mice churning out babies," flowers blooming, a land teeming with fertile flora and fauna, have made him a popular expert on the radioecological zone and have inspired one journalist to imagine a "Garden of Eden just before Adam and Eve showed up."[88] Under the ghostly shadow of Pripyat, the biological research team works in a deceptive "paradise" where "wild boars stomp by, roe deer leap through the waist-high grass and myriad herons and swans feed in the shallows."[89] The zone's wetlands, called "swampland" in early postdisaster accounts, are now "marshland." Walking through the zone with a team of scientists, journalist Alan Weisman found "a broad floodplain covered with meadow grass, daisies, and purple lupine" and teeming with a wide variety of bird species, including "a European goldfinch we'd observed earlier singing in a stand of maples."[90]

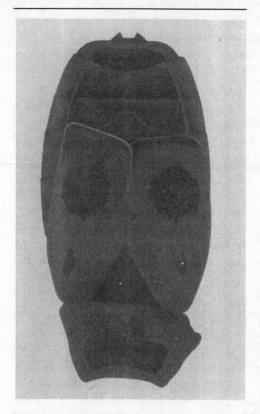

Fire bug (*Pyrrhocoris apterus*)
from Chernobyl. The left side
of the neckplate is deformed
and the black spots are not
symmetrical. (Photo by Cornelia Hesse-Honegger. Courtesy of Cornelia Hesse-
Honegger)

These descriptions evoke a certain ecological discourse, a "language of
engagement" that, according to philosopher David Strong, "invites us to be
fully present in the place."[91] But they seek to expose asymmetries, imbalances,
sicknesses, and weaknesses in the environment, suggesting that their own
pastoral images disguise an invisible menace that at once distorts and ener-
gizes. Solicitations to the reader to appreciate the marsh's science and aes-
thetics, to be sensitive to the environment through which the observer walks,
echo with the ecologist's concern for wilderness preservation. However, the
discourse reaches a limit, for this is a postapocalyptic land where radiation
and other pollutants with unknown properties hide beneath the surface,
outside of any reasonable claims to stewardship. Utterly contaminated, the
land is a "wasteland disguised as bounty"[92] that undermines currently accepted views of nature as useful for colonization, resource development, or aes-
thetic appreciation. Furthermore, forced into transience because of radiation

dosage levels, the zone traveler can never be grounded in such a dangerous place or invent a self that is fully contextualized within this "bioregional narrative."[93] The irony mentioned by many zone observers lies in the struggles of discourse to accommodate the disaster and the incomplete fit of any traditional matrix of descriptive disclosure. The discourse itself is postapocalyptic, heavy with a before and after, unable to find a new language for this unprecedented zone with its vast technological stain, radioactively resistant to disclosure and to mapping, census taking, and monumentalization.[94]

The excitement of disaster's "wilderness" lies in such dangers, in confrontations with what is beyond language and understanding. Transformed into a *terra nova* of alien life forms, the zone provides an analytical site for exploring the observer's own alienation and inability to contain its otherness. In this, I am reminded of Paul Theroux's *O-Zone*, a futuristic novel published in the same year as the Chernobyl accident about the aftermath of a nuclear waste leak that transforms Missouri into a seemingly uninhabitable wilderness. The wealthy "Owners," who live in heavily protected enclaves, journey out to O-Zone in search of thrills and adventure, seeing it as the only remaining earthly frontier and imagining projects that might transform it back into something useful. They are surprised to find people living in O-Zone, and their contact with these "aliens," who were dropped into the radioactive wasteland for crimes against the state, forever dislodges them from their secure lives.

For those journeying to Chernobyl, encounters with residents who remain in the zones provide opportunities for reclaiming the wasteland as an inhabitable site, for locating the compromises that might allow one to live in a poisonous global technosphere. Many people have stayed, or returned to live there, because of strife or danger in other parts of the former Soviet Union. More importantly, the lands near Chernobyl are important cultural sites, where people who fear the loss of their language and tradition have been zoned off.[95] The inhabitants, their culture, and their lands have been resocialized in new scientific narratives of radioactivity and genetic mutation.[96] Organisms take on a new biological meaning as carriers of radionuclides, and the techniques for their domestication are fundamentally altered. Hunting, fishing, and farming now incorporate knowledge of radioactivity detectable only by scientific instrumentation. New food preparation techniques, such as boiling fish in saltwater to leach out radioactive particles, alter daily cultural practices in the inhabited sections of the zone. The human body itself is subjected to continual monitoring: ultrasound screening, whole-body radiation scans, blood tests, electroencephalograms, and tooth enamel sampling.[97] New epidemiologic registries keep track of an affected individual's location, regularly updated medical history, work and other activities at the time of the Chernobyl disaster, and radiation dosage. Ironically, the catastrophic failure of a massive technological system has increased human dependence on complex technologies to the point that the inhabitants cannot live in safety without them. The sense of peril ensures people's cooperation with the new public health regime. If they do not cooperate—if, for example, they choose to carry on their long tradition of gathering mushrooms and berries, now officially declared unsafe—they risk terrible illness.

Despite the promise of this renegotiation with science and its codes and instruments, there are no assurances of good health and longevity, and the overall health status of the affected populations is poor. No one really knows how many deaths resulted from the accident, with estimates ranging from 30 to hundreds of thousands. Discussing the lack of any coherent information regarding the number of casualties, historian David Marples, who has written extensively on the disaster, estimates 100 immediate deaths with minimum total deaths, as of 1996, at around 6000. Anything above, he writes, is "the realm of the unknown."[98] Medical researchers are willing to firmly link only aggressive thyroid cancers with radiation effects,[99] whereas a host of other ailments, such as heart disease, congenital abnormalities, deafness, thrombosis, mental disorders, and impaired immune response are attributed to radiophobia or stress caused by relocation and lack of timely information. As of this writing, a few recent controversial studies have linked chromosomal aberrations and congenital malformations to Chernobyl's radiation effects, with statistically significant rises in leukemia in populations as far away as Greece.[100] Conclusions about Chernobyl's health toll are sketchy for many reasons, including an incomplete understanding of the interactions between chemical and radioactive pollutants (both thrown from the reactor), difficulties in proving a clear cause and effect between radiation and physiological damage, continuing controversies over the effects of low-level radiation, the need for very-long-term studies, inevitable biases in medical studies for political or monetary gain, and bureaucratic mismanagement and secrecy. The overall breakdown of people's health, sometimes known as Chernobyl AIDS, stands witness to the current inadequacy and incompleteness of scientific understanding in a lifeworld radically transformed by disaster. The complex local exchanges among all of the newly organizing entities in the zone are not readily classified or predicted, despite the relentless attempts at monitoring and mapping.

Chernobyl travel narratives often deploy the remaining inhabitants as the voices of brave resistance to this science, maintaining a low-tech agrarian life, described as bucolic: "At dawn, the dark land turns shades of green, yellow and white, and the lumbering shadows turn into dairy cows, many fat with calves."[101] Along with romanticized depictions of animal husbandry and fruit and grain harvesting, the traditional gathering of mushrooms and berries is often mentioned. According to narrative conventions, observers are welcomed into pleasant Ukrainian farmhouses: "Deep in a silent forest, Irina Yashchenko, 53, weeps with joy at the arrival of rare visitors to the ghost town of Szheganka."[102] Here, they are offered homemade foodstuffs by a grandmother who has refused to move and denies any danger. In a poignant moment of contact, the visitors risk the danger of radioactive contamination by sharing this offering: "[Olga] Yakivna pours generous shots of 'samohon,' a potent home brew made from sugar, yeast and water pulled from the well next to her house."[103] Glenn Alan Cheney, during his visit to a "little fairy-tale house" in the 30-kilometer zone, describes how an "old woman serve[d] us some excellent potatoes, beef, slaw, beets, and the best pickles I've ever, ever had."[104] Later, he took the pickles she had given him to a U.S. hospital for analysis, and found their radioactive content to be quite insubstantial. Moscow journalist Masha Gessen describes her refusal to take apples from the Gaia-like Baba Anya, who insists that radiation stimulates vegetation: "We back up into the house in a panic. 'Thank you, but we're afraid,' my friend musters even as Baba Anya fills his camera bag with fruit. Only after we had disposed of the gifts do I learn that, for some reason, apples in the Zone do keep coming up radiation-free."[105] In these accounts, the fearful observer shares a moment of revelation about the radiant land with grandmothers who are nurturing but not reproductive.

Especially before Chernobyl's ten-year anniversary, the residents of the zone appeared as fatalistic old people who believed they had "nothing left to lose"[106] and were tied to traditional agrarian ways of life: "It's my house, my village. It's all familiar . . . I know where everything is. Why should I be afraid?"[107] Their knowledge of their own lifespans sets a limit that outweighs the trauma of disaster. In this, they share the conventions of a very familiar character type in disaster narratives: the elderly man or woman who will not leave spouse or property or pet because of emotional ties. From Ida Strauss, who stood by her husband on the sinking *Titanic*'s deck, to Harry Truman, who stayed on the exploding Mt. Saint Helen's rather than leave his house and his cats, persons who remain deny the disaster and its changes even as they are consumed by them. As one sixty-seven-year-old woman who raises chickens in the zone put it, "I was born here, and I want to die here."[108] The Chernobyl pensioners surface as ambiguities, arbitrating the past with its known focal domestic practices in the milieu of the catastrophic present with its mysterious, foreign, invasive devices: "'All these counters, they lie, they lie,' cr[ies] a querulous group of portly grandmothers."[109]

Other residents emerge as postapocalyptic looters and drifters who carry radioactive furniture out of the zones, and refugees who have no choice but to live in such a catastrophic place. [110] Seeing the matter as a decision between one risk of technological death and another, one refugee from Grozny explained, "You won't find anyone who chooses to live here. We've come here to escape war. Maybe radiation can kill you too, but not as quickly as a bomb. If I had a sackful of dollars, I'd go further. But there's no money, so there's no choice."[111] These voices, speaking of fatalism, resignation, and an apathetic resistance, resonate for other inhabitants of highly polluted lifeworlds. The *Chicago Tribune* placed its Chernobyl anniversary article, about a visit to the zone's "vast ruined land,"[112] below another on a new viral "hot zone" emerging in the United States, a "bestiary of lethal scourges" spreading through overpopulation.[113] As James Rupert explains in his article about the psychological consequences of the "nuclear blight," Chernobyl reveals a "new intimacy of the planet" because radiation sweeps beyond national borders.[114] Interviewed and photographed by a stream of journalists and tourists, zone residents have been inscribed in a globalizing narrative about the terrible normalcy of disaster, the transformations of identity in its disruptions and adaptations, and the inescapability of technological devices and instruments.

Children often figure strongly in these accounts as the most vulnerable victims, whose bodies carry the reproductive legacy of the disaster: "The eyes of children stare unknowingly at a cameraman whose free hand holds a Geiger counter registering off the scale."[115] One visitor to Polessky recounts meeting the four-year-old Katya in the "wasteland" and offering her sweets, speculating on her own possibly deformed future offspring from an "invisible legacy" of radiation: "She walked away from me, the sweets in her hand. Her eyes stared back at me, blinking from under her woolen hat. As we turned and eventually drove away, the image of her pleading, pale features remained with me."[116] These child survivors have been frankly politicized as analytical sites for exploring possible regenerations of faith, especially in medical technologies and the production of consumer goods. In 1996, returning from Kiev, where he visited children in a hospital that treated Chernobyl victims, U.S. Secretary of State Warren Christopher asserted that he had seen firsthand the consequences of Chernobyl and that the United States would help "overcome the poisonous factories, soot-filled skies, and ruined rivers that are one of the bitter legacies of communism."[117] The political rhetoric that the United States is "cleaner" than the former Soviet Union, despite the toxic legacy of its own weapons programs and other forms of environmental devastation, is reinforced in stories about the visiting children from Chernobyl. Sponsored by various humanitarian aid organizations, thousands of Chernobyl children, including former Olympic gymnast Olga Korbut's son, have made the trek to other countries, including Canada, Spain, Ireland,[118] Cuba, Israel, and the United States, where they have been variously deployed as political entities. Like the Hiroshima Maidens who came to the United States in 1955 for reconstructive surgery, the Chernobyl children embody the nuclear holocaust, now a failing logic in the post–cold war climate. They come, it is said, for exposure to a "clean" environment, to "rinse the radioactivity out of their bones."[119] Often described as pale and wan when they arrive, they are restored through consumption of consumer goods: "Families marvel at the speed with which the children become Americanized, quickly developing appetites for everything from two-piece swimsuits to ice cream sundaes."[120] Local papers in the United States feature their Chernobyl children, reluctant to go home, happily playing at amusement parks and dreaming of having their damaged legs repaired so that they can wear high-tech blinking athletic shoes.[121] During their visit to San Francisco under the Jewish Education Center's "Kids Overcoming Katastrophe" program, twelve Chernobyl children spent their free time visiting Disneyland, playing on donated networked computers, and "soaking up fresh air, pure water, non-poisonous food and all the free-wheeling American culture they [could] absorb in a year."[122] Perhaps the oddest donation to the cause of Chernobyl children is the sperm bank in Kiev, contracted in 1996 with the New England Cryogenics Center. Its purpose is to provide frozen American sperm to reinvigorate the birth rate of the Ukrainian populace, since nearly half of thirteen- to twenty-nine-year-old men there have fertility problems.[123]

While the zone residents eerily soak up a hidden radiation from plants and soil, the Chernobyl children, in these images of culture transfer, are cleansed and reinvigorated by the contaminations of culture. At the same time, a cultural imaginary, nationalistic in character, is inscribed on their bodies, warding off the disaster through its appropriation of them: "[Katja] had great skin color and had gained some weight. And when her father saw her again, he said Katja now looked like a little Canadian girl."[124] Although the children may indeed physically improve from trips abroad, the real beneficiaries are their benefactors, who confirm the safety of their own lands and nations, absorbing the children's otherness as radiation victims and containing them in philanthropic narratives. Even when these children are presented as voices against nuclear energy development, it is with the thought that the disaster has not yet happened here and can yet be prevented by welcoming its refugees.

After the catastrophe that "chopped them in half,"[125] survivors negotiate new codes of identity produced by external observers. Often, they are well aware of their deployment in these scenarios. Discussions of their physical and psychological condition fold back into the victims' lives, giving them new terms for describing their condition, a language for survival in a world where scientific knowledge of radiation effects and access to medical interventions have become essential. That language is a baroque menagerie of diseases. A Chernobyl liquidator explained: "Our complicated medical problems need complex medical treatment. I do not remember a single day in the past two years when I have felt well. What can I do when my medical card repeatedly confirms: second grade encephalopathy, dolichostigma, gastritis, cholecystitis, chronic prostatitis, haemorrhoids, functional problems of left heart ventricle and other 'minor' conditions."[126] Because of the long-lasting nature of radiation effects and the slow arrival of catastrophic illnesses, this writing of the disaster does not attempt to contain or deny death, but to reveal it. At the same time that the voice announces its tenacious survival, it speaks of death and the failures of that discourse, for as another liquidator says, "We are dying, there are less and less of us, the Chernobyl disaster liquidators. Medicine cannot save us."[127] While the makers of disaster firmly announce its survivability, survivors everywhere are caught between these worlds of the before and after, in the half-life of the uncanny, uncertain, catastrophic present.

Alla Yaroshinskaya, a Ukrainian journalist, fled with her children from her own town, Zhitomir, and then returned to investigate the relocation of Chernobyl refugees to hastily built towns on equally radioactive lands in Narodichi district. She writes of this work, "I have kept the notebooks I filled at that time, and they sweat blood, like the memories of people who have reached a great age and who no longer expect anything from life."[128] Unlike medical records and official reports that dismiss illness as a psychological influence of their own external discourse, Yaroshinskaya suggests that her notes have a closer entanglement with the bodies of the victims, that they in fact speak out of, rather than into, those bodies. For survivors of Chernobyl, the transcendence into nationalist rhetorics, medical diagnoses, scientific reports, and travelogues is remote from the painful uncertainties of the transformed world in which discourse is intimate, but too often fails. Yaroshinskaya describes seeing a boy from Staroye Sharnye running out into a garden near the fence to the Forbidden Zone, carrying signs that warn of death: "He runs joyfully into the middle of the cascading autumn flowers. He is so sweet, he looks like a flower himself. His little hand reaches out towards the apple, he brings it to his mouth . . . the juice runs over his lips, falls on his shoes, the grass."[129] Such a passage speaks of possible misreadings, not only in the boy who is oblivious to the official signs, but in the survivors who must read beyond what Yaroshinskaya calls a "real picture" that evokes "filial affection" to the radioactive hell behind all the narrative surfaces, including her own. To stay in such a place requires holding on to a traditional reading, denying the new undertows in a well-known, mythic landscape, even at the risk of one's own life. These split readings, caught between denial and acceptance, are ultimately uninhabitable, but they are now familiar to the many living in the radioactive, polluted landscapes of the former cold war, where the disaster is not a monument, but a way of life. That radiation effects spread far beyond any national borders suggests that far more people are engaged in these kinds of readings than the once and present residents of Chernobyl.

3. Radioactive Body Politics

Ground Zero

Since the atomic bombing of Hiroshima, images of nuclear holocaust have proliferated in cultural discourse in the United States, marking every large-scale disaster. In these nuclear narratives, horrific destruction spreads in concentric rings from Ground Zero, the epicenter of the blast, and engulfs the unsuspecting inhabitants of surrounding communities in a blastwave of death. The postapocalyptic world is so racked with pain and loss that the numbed survivors envy the dead. Our understanding of AIDS has been shaped in part by pervasive references to this nuclear text, a text that has particular meaning in the work of AIDS activists writing in the 1980s, amid other narratives on the fragility of a prosthetic world designed for safety. For these writers, the conventions of nuclear protest provided a way for entering political discourse as well as a shared language for exploring grief, loss of community, and survival.

The fictional survivor of a nuclear holocaust bears the burden of restoring community, culture, and human continuity from the ashes.

That is, if he is not the last man standing. The last man is a familiar figure in nuclear discourse, crawling from his shelter to find the world in ruins, free at last of the constraints of social mores. A most famous last man appeared in a *Twilight Zone* episode starring Burgess Meredith as an ineffectual bank teller with a passion for reading, unable to do so at home because his wife cruelly scribbled in his books.[1] He hid in the bank vault to read on his lunchtime, until one day, he emerged into a world devastated by nuclear holocaust. Moving through the rubble of torn couches, broken tables, and twisted beams in stunned despair, he suddenly realized that he was now free to read all he wanted, and went to the library to immerse himself in the Western canon of great books. On the steps of the library with its great clock, he dropped his glasses, they shattered, and he realized that he had nothing left, nothing left but time. Carrying the hope of cultural continuity, the archive finally failed the last man, robbed of sight by a small, final technological disaster. There was no one left with whom to communicate. Words failed.

In his collection of essays on the impact of AIDS on the urban gay community, *Ground Zero*, published in 1988, Andrew Holleran presents himself as the last man, the survivor of an urban holocaust for whom words fail. He adopts some conventions of nuclear fiction to describe New York City after the advent of AIDS—after "the bomb fell."[2] Holleran explores the futility of representing the AIDS landscape and the ironies of turning the X-ray of a textual examination onto the survivors. After what Paula Treichler calls "an epidemic of signification,"[3] what could one possibly say about AIDS and who was left to listen? Like Treichler, Holleran presents AIDS as a construction of language: "Many of us have seen our generation wiped out in announcements from the Harvard School of Public Health. Some of us have been told we were terminally ill by Barbara Walters."[4] AIDS was socially constructed with such pronouncements of doom, inflated by a rhetoric of atomic urban destruction. Thus, Holleran suggests that reading about AIDS was like "staring at the sun,"[5] an image that, given the book's title, evokes the atomic bomb, brighter than a thousand suns. The AIDS text produces a kind of flash blindness, an inability to gaze upon the face of disaster.

Aware of his own difficulties as a novelist and journalist in the age of AIDS, Holleran explains, "Writers who dealt with homosexual life before the plague— the manners and mores of the homosexual community—have been quite left behind by a change of circumstances that blew the roof off the house they had been living and writing in."[6] This catastrophic explosion transformed the body into "diseased meat," an "object of dread" while survivors bargained with death. Holleran's role, in the midst of the plague, is to record those bargains and assess good or bad behavior. Holleran, as the angel of judgment who pos-

sesses the final book of names, has to abandon sex, the body, the reproduction and renewal of gay urban culture. Instead, he has to find his purpose in
remembering, preserving, and creating an archive—an archive that would
stand against a final collapse of meaning.

Holleran's venue is the gay urban institutions of New York, from the bath-
houses to the penthouses—now in ruins and haunted by ghosts. In his title
essay, he describes the transformation of the Metropolitan Theatre from a
"dark, warm chamber of seed" to a place brightened and exposed by "light,
law, manners, reality," where sperm lay on the floor "like plutonium."[7] Holleran's own text is responsible for that illumination: an X-ray of the postapocalyptic sex theater with its isolated denizens. The author wanders detached through this theater, observing the numbed survivors: "They wander
back and forth—across the foyer, up and down the aisles, searching, searching."[8] But the author can never escape his own identification with them.

This scene is reminiscent of survivors' memories of atomic bombing of
Hiroshima and Nagasaki, often represented in nuclear protest literature during the cold war years. As one survivor recounts, "Wherever I walked I met
these people. . . . Many of them died along the road—I can still picture them
in my mind—like walking ghosts. . . . They didn't look like people of this
world. . . . They had a special way of walking—very slowly. . . . I myself was
one of them."[9] Robert J. Lifton calls this postdisaster experience a "death in
life" for the survivors, whose complete immersion in wholesale slaughter
marked them forever, physically and psychologically, as the *hibakusha*. To be
hibakusha meant to have constant lingering fears that "A-bomb sickness"
would manifest itself so that the body had to be constantly monitored for
symptoms—fatigue, spots, stomach trouble.[10] To be *hibakusha* meant a complete transformation of identity, a public marking with the death sign of the
Bomb. This experience of the walking dead was, according to Lifton, continuous, interminable, and unresolvable.

Through interviews and psychological and medical analyses, nuclear protesters attempted to reinscribe the *hibakusha*'s "human truth" within a statistical language of megatonnage, crater depth, body counts, percentages of
buildings destroyed, rads, and blastwave measurements. The scientific study
of mathematically describable effects was applied to cities in the United States
so that readers of nuclear protest literature were expected to see themselves
mirrored in the bodies of Hiroshima victims, to see their own bodies drawn
within the chalk outlines on Hiroshima's pavement.

The *hibakusha* presented an alternate narrative to the linear unfolding of
technological apocalypse. The representation of nuclear bomb survivors—
from the Hiroshima Maidens to the slowly dying farmers wandering the ru-

ins of Kansas City in the 1983 television film *The Day After*—sought not only to construct an alternative to the cold, rational language of strategic discourse but to urge identification and action. In the 1980s, nuclear protesters began to call for nothing less than a transformation of human consciousness and community, from Jonathan Schell's urgent, "Let us connect," to Helen Caldicott's, "We are all sons and daughters of God, and under the universal horror of the Sword of Damocles, we will be united together in mutual respect and peace."[11] Literary scholars called for a "nuclear criticism" of cultural texts and artifacts that would expose "unconscious nuclear fears," and reinvent human community.[12] Indeed, the conventions of nuclear protest writing were established as a representation of a nuclear weapon's explosion and immediate effects on the landscape (vaporization, blastwaves, firestorms), a description of the physical and psychological effects of the nuclear blast and radiation sickness on human bodies, and an evocation of survival through a transcendent and psychologically transparent collectivity.

Because of the popularity of writers such as Caldicott, Lifton, and Schell, and the proliferation of nuclear scenarios in science fiction and television films such as *The Day After*, the conventions of nuclear protest became entrenched in U.S. culture, especially in the late 1970s and early 1980s. Consequently, they provided a model for other protests against cultural inertia in the face of disaster, including AIDS. Two threads of nuclear discourse emerged together in AIDS activism: a call for an atomic war to be waged at the virus and the representation of HIV itself as the apocalyptic agent within. The first borrowed the familiar language of military strategy that could be shared by other political constituents and biomedical researchers looking for federal funding. In the allocation of federal monies, a Manhattan Project for AIDS could replace the absurdly baroque Strategic Defense Initiative. Even old enemies such as Larry Kramer and Cardinal John O'Connor could agree that a Manhattan Project for AIDS was needed.[13] In one of his frequent iterations of the theme, Kramer told *Playboy*, "What we need is a Manhattan Project—like the group of people in the Forties that the government sent into the desert to build a fucking bomb. They were thought to be the best we had. 'Here's the money to do it, here's the staff and don't come back until you do it.'"[14] In the same vein, Emmanuel Dreuilhe wrote, "When will another Oppenheimer discover the fearsome weapon that will resolve this internal dissension and sweep the virus from the face of the earth?"[15] With his coauthor, Mauro F. Guillén, Charles Perrow, who theorized technological disaster for the 1980s, called for an AIDS initiative based on the 1950s Polaris missile program that prospered because of patriotic fervor, abundant funding, and relaxed scientific standards.[16] Michael Sherry argues that the

1980s "war on AIDS" rhetoric—including references to the holocaust and nuclear apocalypse—urged collective action through the expression of "a modern liberal faith in concerted federal action," but that such expression was short-lived and ambiguous.[17] During his 1992 campaign, Bill Clinton promised AIDS activists a Manhattan Project for AIDS, but that effort never materialized. In the biomedical community, the image of a Manhattan Project for AIDS has persisted, but has been neutralized of its political fervor. Various groups have been called the AIDS Manhattan Project, including HIV discoverer Luc Montagnier's World Foundation for AIDS Research and Prevention, the Office of AIDS Research under the National Institutes of Health (despite the director's objections to the language), the AIDS program at the University of Alabama at Birmingham, and the Aaron Diamond AIDS Research Center in New York. Now, researchers refer to these centers as "Manhattan Projects" mostly to describe the concentration of brilliant young talents devoted to a single problem and sharing a noncompetitive comradeship.[18]

The early opportunistic rhetoric of a nuclear war on AIDS was accompanied by another call to collective action, shaped by the conventions of nuclear protest, a transformation of consciousness catalyzed in the experiences of those who had suffered a viral apocalypse. Early on, diverse political constituents shared the aggressive strategic language of the war on AIDS, but these nuclear protest conventions were the domain primarily of AIDS activists interested in reshaping identity and community. As Richard Dellamora points out, gay writers were "pressed into service as angels of the millennium," using the image of nuclear disaster "to signify unspecified anxieties about continued individual and group existence in the face of AIDS."[19] Literary critic William Scheick defines nuclear criticism as "a re-minding, a hoped-for reinterpretation of communal memory . . . that would contribute to a revision of human consciousness."[20] In this tradition, AIDS activists called for a much broader transformation of community, well beyond the strategizing of prevention and cure. Such a stance presupposed that a utopian, transcendent community, centered in a domestic "life affirmation" against the broadly painted political forces of "death," was possible and desirable. Thus, in his last essay in *Ground Zero*, Holleran, rejecting his own silent "last man" status and affirming his emotional ties to those living and dead, writes, "The fact that people die does not mean we stop talking to them. It may mean we start talking to them. Especially when the people who have been left behind feel guilty about the fact; baffled by the accident of their own survival."[21] Even the dead could be kept close against the disaster in this new vision of community.

Patient Zero

The use of nuclear protest conventions was strikingly evident in Randy Shilts's *And the Band Played On*, a history of the early AIDS crisis, later made into a television movie.[22] The book not only makes reference to the doomed passengers on the *Titanic*, but reflects the nuclear scenario of escalating, concentric crisis, a scenario that was firmly planted in cultural discourse during the cold war. Shilts writes that medical researchers "marked the spread of AIDS in concentric circles, pulsing out of the center of Manhattan to include larger and larger rings of land and population in the impoverished outlands of metropolitan New York City."[23] The structure of *And the Band Played On* carries this apocalyptic topology of a radiating menace. Peter Schwenger suggests that nuclear narratives revolve around an epicenter, a textual ground zero, a paradoxical absence suggesting a mathematically infinite regression.[24] Schwenger analyzes Russell Hoban's postapocalyptic novel *Riddley Walker* in terms of the absent center, concluding that the text is postmodern bricolage, a playful journey of riddling and unriddling that radiates from the unthinkable zero point.[25]

Shilts's AIDS narrative also circles around disaster's absent center—the elusive original bearer of the virus, the hypothetical Patient Zero. Patient Zero has several incarnations in the book, but he was most obviously Gaetan Dugas, an airline steward who supposedly spread HIV across the Western hemisphere. Epidemiological tracings of the virus have perpetuated the idea that transportation technology is transforming people into conduits for a biological apocalypse, from tourists on airplanes to Indonesian truck drivers. Just as airplanes transported nuclear bombs, they carried the infected Dugas and thus "cast the seeds of the apocalypse."[26] Shilts first introduces Dugas as a participant in the 1980 San Francisco Gay Freedom Day Parade. Death was also marching, "elbowing its way through the crowds on that sunny morning, like a rude tourist angling for the lead spot in the parade."[27] Clearly, Death and Dugas were one—Dugas was a tourist from Toronto, a highly competitive person who was always "the prettiest one."[28] The figure of Dugas is a metaphor for the epidemic, dividing lives into before and after the zero hour, "a commonly understood point of reference around which an entire society defines itself."[29]

Patient Zero was an unwitting recorder of the epidemic who carried matchbooks and napkins filled with the names and numbers of potential and actual lovers. He also had an address book inscribed with his "most passionate admirers," who were often fleeting memories: "At times, Gaetan would study his address book with genuine curiosity, trying to recall who this or that person was."[30] In Shilts's narrative, this act of writing—this postmodern bricolage of desire emanating from an infinitely seductive Zero—was without meaning to Dugas, who was pictured as scarcely remembering his own authorship of these addresses. It reflected the spectacular 2500 lovers of Patient Zero, an inflationary sexual surplus. In keeping with the high moral tone of apocalyptic discourse, Shilts rails against this depersonalization and commercialization of sex, an anonymous sex, a text without clear authorship, a list of names without significance.

Patient Zero's collection of names and addresses, of random scraps and scrawls, became meaningful only as a social record of infection, and only when interpreted by the Centers for Disease Control (CDC), which established the trace of the virus spreading from Patient Zero. They calculated "the odds on whether it could be coincidental that 40 of the first 248 gay men to get [Gay-Related Immune Deficiency] might all have had sex either with the same man or with men sexually linked to him. The statistician figured that the chance did not approach zero—it was zero."[31] Thus, HIV retrospectively endowed Patient Zero's chance encounters with a formal, mathematical structure, subject to a rational hermeneutics. Carried in the airline steward, the virus created a "permanent demarcation" between before and after, cleaving lives "in two."[32] These references to a sharp rupture in lives and times evoke fission, a splitting that released an unstoppable energy, a principle of plenitude, a mathematical logic of AIDS. Shilts's body count reinforces this sense of escalation, interrupting the narrative at regular intervals.

Thus, Patient Zero's names were given a collective, public meaning as they were marked with HIV. Whereas Jacques Derrida calls nuclear war the Apocalypse of the Name, AIDS activists figure AIDS as the Apocalypse of the Names. The work of the remainder, the reading of the Names among AIDS activists, ceremonializes this epidemiological marking and multiplication while resisting the total logic of destruction that would eliminate all naming. Like the *hibakusha* who came to Hiroshima's Peace Park to place the escalating names of the dead in a cenotaph, or Bhopal's survivors who placed stones bearing the names of the dead within Union Carbide's factory walls, AIDS activists accumulated the names, marking them on quilt squares and banners and reading them off in public demonstrations. These events preserved the names and their symbolic content against the transitory statistical nature of epidemiological tracing that marked the names for death's erasure. But in *And the Band Played On*, the rhetorical strategy is both to preserve the names and to contain them within a moralizing eschatological discourse.

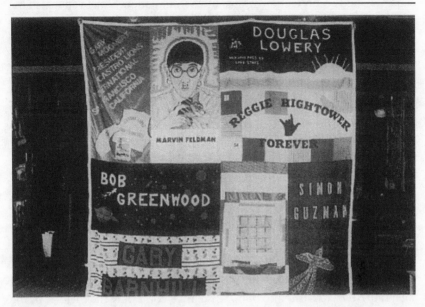

A Quilt section. (Photo by Names Project/David Alosi and Ron Vak)

Reflecting civil defense maps of Ground Zero, *And the Band Played On* is a set of concentric interpretive circles: Patient Zero's collection of names was contained within the CDC's epidemiological interpretation, which was contained within a metanarrative about the efficacy of social and medical solutions to the epidemic. Thus, the text itself is an effort to contain AIDS in a politicized construction of language by diplomatically but strategically maneuvering language to contain the explosion of names. Shilts charges the government health bureaucracy, the press, and the gay urban community itself with allowing the virus to spread, but Shilts provides retrospective solutions that seem oddly nostalgic. Sexual freedom was replaced by harmonious domesticity. "Meaningful commitment" stood opposed to "omnipresent carnality." Shilts writes that by 1985 "more than half of gay men were ensconced in relationships,"[33] sheltered, covered, secure, safe from Patient Zero's fallout. As Gillian Brown writes, the appeal to the domestic has an intimate history in nuclear rhetoric, from fallout shelter manuals that preached family harmony underground to women's peace movements that used the iconography of baking, gardening, and weaving.[34] The domestic space represents continuity, sequence, and accumulation against utter annihilation. Shilts's appeal to the domestic, reflected also in The Names Project Quilt, steals from the family values toolbox of the religious right, but also evokes the spirit of "Bread, not Bombs." In the theory of nuclear protest, a highly romanticized domestic iconography was supposed to break down the separation of people from one another and allow them to commit to a meaningful, life-sustaining collective purpose.

Within *And the Band Played On*'s narrative structure, its fallout shelter of words, Shilts constructs the most intimate thoughts of his subjects, a strategy that results in a metanarrative of a transcendent collectivity, a utopian vision of sympathetic beings who, caught in the purifying fire of AIDS, were transparent to each other. The book closes with a spiritual vision of connection that ended the blastwave of AIDS statistics. A nurse came to an understanding of life while contemplating the body of Bill, who died of AIDS: "'But that's not him. . . . He's not really dead.'"[35] Bill's death lifted those around him above the holocaust into a new realm of sympathy and enlightenment. Like the language and form of nuclear protest, a traditional Christian eschatology helped integrate AIDS activism into public discourse, contesting the rhetoric of the religious right while calling for the separation of saints and sinners. Andrew Holleran, for example, suggests that writing about AIDS will end finally in a book of judgment, "a simple list of names—of those who behaved well, and those who behaved badly, during a trying time."[36] Thus, Patient Zero's names were marked and charted by moral imperatives, pastoral letters.

Tony Kushner, in his two-part play *Angels in America*, suggests that this use of apocalyptic iconography is quite theatrical and even clumsy, even as it offers some communally constructed meaning.[37] Like *And the Band Played On, Angels in America* allows its characters to read each other's minds, to slip in and out of each other's daydreams and fantasies. But this extrasensory perception, apparently a gift from the avenging angels of the approaching millennium, is accompanied not by calm reflection but by a jarring sense of invasion and dislocation in time and space. In the play, an AIDS patient's dream about applying lipstick blends with a housewife's Valium-induced hallucinations, and Roy Cohn receives a visit from Ethel Rosenberg. The recurring references to hurricanes, explosions, holes in the ozone, sinking ships, raging fires, medieval plagues, and most of all AIDS add to this sense of disorientation and apocalyptic doom.

The heavenly signs that signal the imminent End are most emphatically theatrical illusion; Kushner advises that the director let the theatrical wires show. These signs include spectacular lighting effects, a miraculous book that opens to reveal a flaming Aleph, and an angel who descends like a missile, crashing through the roof of a house, provoking its AIDS-afflicted resident to cry out, "Very Steven Spielberg!"[38] Theatrical effects reflect the medical technologies surrounding and penetrating the bodies of patients that give only the illusion of ultimate judgment and salvation. The pentamidine IV drip is as much theater as the angel's Book of Life. Nor can technological mastery provide security against the End; Roy Cohn's virtuoso performance on his phone system cannot prevent his demise. *Angels in America* ends with the words, "The great work begins," suggesting uncertainty, expectancy, possibility, events unfolding outside of theatrical effects and conventions that attempt to contain a destabilizing force. The name of *Angels in America*'s second half, *Perestroika*, suggests the collapse of the cold war's fixed order, in itself a fabulous theatrical effect that made the careers of the Bosses, the defense planners and atomic spy hunters such as Roy Cohn. After the destruction of the cold war's strategic balance, Kushner suggests, the great work was living affirmatively and with mercy in an imminent, ever-unfolding disaster that always fell short of the big End.

Robotic Kid

The threat of viruses and other diseases that might assume an apocalyptic agency in the geographic movements of people and animals also gave rise to cultural fantasies about bodies made invulnerable through technological enhancements. HIV, increasingly associated with a menagerie of viruses supposedly emerging from the dreaded "hot zone,"[39] seemed a failure of both institutional and physiological barriers. Stemming from military research and development, cybernetics offered a vision of adaptable bodies that could survive in hostile environments with dangerous agents such as toxic chemicals and radioactive particles. During the height of the cold war, the U.S. military funded the development of teleoperators, mechanical exoskeletons for soldiers that would provide them with bodily protection and superhuman strength. Like the fallout shelter, the HARDI-MAN was expressly designed for shelter against a radioactive environment.[40] But the HARDI-MAN allowed for movement, interaction, and manipulation, and suggested that humans and machines could be joined in an emerging symbiosis. In the 1960s and 1970s, cyborgs entered biomedical discourse on the possibilities of organ transplants and prosthetic devices, and popular culture in the figures of the Bionic Man and Bionic Woman.[41] The ultimate hope of bionics was that the body would grow stronger and less prone to diseases, accidents, and the ravages of war.

In the 1980s, the cyborg emerged again as a being that could withstand any disaster, even the ultimate one. The popularity of Stephen Hawking, stemming from the 1988 publication of his bestselling *A Brief History of Time*, revolves around his survival of personal disaster, the onset of amyotrophic lateral sclerosis, and his mediation of a much larger disaster, the black hole, a star collapsing under the weight of its own gravity, the gaping maw that could swallow the universe, familiar in popular accounts of astrophysical research in the 1980s. At the beginning of Hawking's popularity, the media formed a set of conventions for describing him that made his work a metaphor for his own body. He was physically collapsing into himself as though into a black hole: "Like light from a collapsing star, exhausted by the struggle against gravity, the thoughts of Stephen Hawking reach us as if from a vast distance, a quantum at a time."[42] His biographers write that Hawking appeared in popular science writing of the mid-seventies as "the black hole cosmonaut trapped in a crippled body, piercing the mysteries of the Universe with the mind of a latter-day Einstein, going where only angels feared to tread."[43] As he described to his popular audience an astronaut extruded into spaghetti after passing the lip of a black hole, Hawking presented himself as a witness to the ultimate disaster from which he always returned.

The spectacle of Hawking's body—and never has such attention been paid to the body of a physicist—has been his waning ability to speak and his increasing dependence on machines. Writing in *Film Quarterly*, Shawn Rosenheim observes that Errol Morris's *A Brief History of Time* (1992), a film about Hawking's life and work, borrows heavily from *The Incredible Shrinking Man* (1957), whose protagonist begins to dwindle to nothing after exposure to atomic fallout. The film's narrative and editing compare Hawking and black holes as objects of similar study, intimating that "Hawking will likely suffer a kind of information implosion, leaving him unable to signal to the world by any means; he will be a silent object, unable to confirm or deny his own interior experience."[44]

Hawking has been saved from this so far, as his popular biographies go, because computerized voice technologies keep him projected into the world. In her book on gender, technology, and cyberspace, performance artist and cyber diva Allucquere Rosanne Stone describes witnessing a lecture by Stephen Hawking, who talked through his computer system, outfitted with a speech synthesizer. Meditating on the intersections between bodies and technologies, Stone observes that Hawking's identity extends beyond his physical body, losing the borders of what we normally see as a person. But without the machines on which he depends, she argues, Hawking's "intellect becomes a tree falling in the forest with nobody around to hear it."[45] Hawking's status as a cultural icon rests on a disaster narrative involving potential loss of work, movement, language, and sexuality, the connectivities in assemblages of humans and machines. That Hawking can speak through his integration in a cybernetic system reverses the elements of the 1980s disaster narrative, in which fractured technological systems drive living beings into death, sterility, unsustainable mutation, and the silence of pain.

In the cultural imaginary of Hawking, his lifeboat of a sort is the technological system that provides escape from a natural disaster that has the same effects. However, the complete apparatus, such as Hawking's breathing mechanism, is not the focal point for analysis, but his computer-synthesized voice, run by his laptop. Cybernetics provides the avenue for sustainable interior work, carefully dictated to the outside through a tight biotechnical interface, as in Biosphere 2. However, as a model for information workers, increasingly dependent on computer transmissions themselves, Hawking's image is fraught with anxiety. The media's attention to Hawking's sexuality, his marriage to his nurse, the pictures of Marilyn Monroe in his office, his *Playboy* interview, his statement that he'd sold more books than Madonna, his bet with physicist Kip Thorne that he'd give him a *Penthouse* if Cygnus X-1 was ever proven to be a black hole,[46] all assuage the fear that the body—especially the male heterosexual body—might be lost in some terrible disaster of transcendence.[47]

The much lesser known David Wojnarowicz, an artist and writer who died of AIDS in 1992, adopted the persona of the biomedical cyborg to disrupt the power relations embedded in biotechnical systems. Unlike the stabilizing narratives surrounding Hawking, Wojnarowicz described himself as "a bundle of contradictions that shift constantly."[48] Whereas AIDS activists embraced the stable order of the domestic shelter in public discourse, Wojnarowicz borrowed instead from William Burroughs's technologically enhanced Wild Boys, terrorists and saboteurs, armed with "laser guns, infra-sound installations, Deadly Orgone Radiation"[49] stolen from the police states of the Western world. Rather than constructing AIDS in the nuclear protest tradition of peaceful hearth, Wojnarowicz used the tough, mutant persona of a character from a postapocalyptic novel, roaming the barren countryside, using any means to survive, and chronicling his own political resistance and physical disintegration. In his artwork, he often drew on medical and nuclear imagery, juxtaposing photographs of battlefields, gay sex, cells seen through a microscope, and houses exploding during nuclear bomb tests. Art critic John Carlin describes Wojnarowicz's work as an "evocative poetry" that is both "apocalyptic and transcendent."[50]

In a meditation on sex, nuclear technology, and artistic vision, Wojnarowicz wrote, "We are born into a preinvented existence within a tribal nation of zombies."[51] Under that illusion of tribal oneness, the many real tribes had various investments in power, including the telecommunications tribe, which fatally lulled its viewers with offers of hope. Only "stray dogs" could escape the zombie world through intelligence, experience, language, and new technologically enhanced vision, an "X-ray of Civilization" that would expose and dismantle cultural foundations and structures.[52] After a trip to an atomic bomb museum where he encountered the uncanny neutrality of soundtracks and educational films and tourists snapping photographs of incineration and death, Wojnarowicz experienced his own moment of separation from the "thick and hallucinogenic" productions of the "illusory tribe" when he had sex with a stranger in his car. Wojnarowicz found himself fissioned, transformed into the haywire "robotic kid looking through digital eyes past the windshield into the preinvented world."[53] These eyes were capable of scanning through scale and proportion, freezing images and penetrating bodies. Thus, Wojnarowicz stole from the image of the militaristic rogue cyborg who was unaffected by a hostile, radioactive environment, as in films such as *Cyborg, Universal Soldier,* and *Terminator.*

Rather than constructing himself in the media image of the sainted, devastatingly wasted AIDS victim who was supposed to remind us that we can spiritually transcend physical disintegration and mortality, Wojnarowicz took the hacker's role of technological disruption and disturbance on the "outskirts of tribal boundaries," his eyes and limbs prosthetically enhanced: "Surveying the scene I wonder: What can these feet level? What can these feet pound and flatten? What can these hands raise?"[54] As Andrew Ross writes, the hacker's knowledge penetrates and rewrites cultural programs based on systems logic and opens the way for alternative uses of technology.[55] As a viral cyberjockey, Wojnarowicz navigated the codes of the culture's soft machine, resisting a biomedical model of the objectified HIV-infected body and escaping the "flotation tank" and the "dry, rug-covered terrarium."[56] He became the technologies that oppressed him—the microscope, the X-ray, the computer. As Felix Guattari writes "The authenticity of his work on the imaginary plane is quite exceptional. His 'method' consists in using his fantasies and above all his dreams, which he tape-records or writes down systematically in order to forge himself a language and a cartography enabling him at all times to reconstruct his own existence. It is from here that the extraordinary vigor of his work lies."[57]

Medical technologies contain bodies, forcing them into passivity. Whether sitting for hours waiting for the AIDS specialist or lying motionless in a coffin-like metal tube for magnetic resonance imaging, people with AIDS spend a good deal of their time trapped in an oppressive stillness, surrounded by medical technologies. Since E. M. Forster's classic story "The Machine Stops," fear of the machine's enclosure has been represented in human bodies turned into fleshy lumps, stripped of their limbs and sensory organs. Stories about nuclear holocaust often contain such imagery. In Judith Merrill's "That Only a Mother," a fetus subjected to post-Bomb radiation is born without arms and legs; in Daniel Galouye's *Dark Universe*, remnants of the human race navigate through the dark corridors of their vast underground blast shelter without eyes; in Harlan Ellison's "I Have No Mouth, and I Must Scream," an omnicidal computer, which destroyed the human race through nuclear war, tortures the lone survivor, transforming him into an eyeless, armless, legless, and mouthless slug.[58] These tales suggest that the technological systems that promise safe enclosure strip away our ability to move, to perceive, to act, to make.

The long-term survival of people with AIDS may depend on their ability to resist this passive model and "act up." Resistance to homophobia, sexism, racism, and general ignorance in social services and the health care system helps people with AIDS stay engaged and active. Many people with AIDS have learned to access medical information and navigate its codes. Through information sharing, they are able to develop and access alternative funding sources. Furthermore, as Paula Treichler and Lisa Cartwright suggest, AIDS activists have attained the "media savvy and technological sophistication" to counter the popular media spectacle of AIDS with their own inscriptions of identity.[59] As the work of Wojnarowicz suggests, survival—and not mere survival—means being fully projected, differentiated, and sensitized within the symbolic exchange, resisting the collapse of the body into the zero point of viral or nuclear holocaust.

4. Oil and Water

Driving under the Influence

In the media spectacle of the *Exxon Valdez*, the drama unfolded along an easily recognizable opposition between nature and technology. In this theater of disaster, bodies were positioned according to their traditional relations with certain technologies and expected to perform gestures according to class, gender, and race. The media trotted out a cast of players, from the drunken captain, Joseph Hazelwood, to the mostly female bird and otter rescue workers, presenting various moral lessons for life in the oil-fueled technosphere. Exxon managers appeared as "fishoil" salesmen preying on innocent locals: brave-hearted fishermen, Edenic Natives, and otter mothers with pups. And complex corporate technologies were pitted against domestic and subsistence technologies in visions of survival in hostile environments.

Hazelwood was the first figure to emerge in the spectacle: "Joe Schmoe, a boat driver,"[1] as he described himself, who had suddenly become a national villain. Six years later, he would be perversely canonized in *Waterworld,* the ultimate postapocalyptic lifeboat film, in which evil, oil-consuming, hard-drinking "Smokers" consider Hazelwood a saint as they ply a rusty *Exxon Valdez* and murder the environmentalist sailboat dwellers.[2] About ten hours after the accident, the Coast Guard delivered a blood-alcohol test that showed Hazelwood's count to be 0.06, above the legal level of 0.04 for marine vessel operators. The media then exposed Hazelwood's DUI convictions, for which he had lost his driver's license three times. At the time of the spill, Hazelwood's driver's license had been revoked. Initially, before it appeared that the blame might boomerang, Exxon managers went along with the media focus on Hazelwood's drinking as the cause of the disaster.[3] Even though Hazelwood was not piloting the ship at the time of the collision, the image of a drunk driver who couldn't get through a waterway through which "your children could drive a tanker"[4] persisted. Hazelwood's own lawyer, Michael Chalos, called his criminal trial "the biggest drunk-driving case ever put on."[5] In his closing arguments at the criminal trial, prosecutor Brent Cole compared Hazelwood to "the young kid who's driving down the street in kind of a souped up car and he's going too fast through traffic."[6]

A drunk driver in an unsafe vehicle, Hazelwood represented for environmentalists the messy, hallucinatory intoxications of Big Oil, an "infernal combustion engine"[7] invading a pristine wilderness, the machine in the garden that suddenly ruptured. Art Davidson, natural resource planning director for the State of Alaska, found the *Exxon Valdez* oil spill to be "a story of addictions: not just a tank captain's addiction to alcohol but widespread addictions to power, money, and energy consumption."[8] Other environmentalists sounded this broad theme of national addiction and driving under the influence of oil companies: "We cannot point the finger of blame for the *Exxon Valdez* when the underlying cause is a country completely drunk on oil."[9] In his book on the disaster, John Keeble writes that to ensure U.S. affluence, "we require stability at the wellhead so that we may bend to it and drink deeply of the dark stuff while all around its edges the garden erupts with fire."[10] And some Alaskans themselves spoke of their economic relations with Big Oil in the same terms. Alaska fisherman Tom Copeland told Davidson, "Alaskans are addicted to that oil money. We've got that needle in our arms."[11] Dave Cline, director of the Alaskan National Audubon Society said that Hazelwood "was just a captain of a ship trying to get the oil to market and put in this position because of America's gluttonous appetite for oil."[12] The theme of all U.S. citizens' culpability for the disaster was further enhanced by the frequent use of the automobile and its emissions as a metaphor for the *Exxon Valdez.* The image not only indicted oil-addicted, gas-consuming car owners, but also normalized the disaster in a discourse on everyday high-risk technologies.

Although Hazelwood was cleared in an Alaska court of the criminal charges of driving while intoxicated, criminal mischief, and reckless endanger-
ment,[13] his public image spoke of this addiction to danger in particularly
visceral, scatological ways. Following his criminal trial, Hazelwood told a local
reporter that the impact of the ship "was like a big punch in the gut . . . like
the wind is knocked out of you, but there's no recovery."[14] In his testimony
at the subsequent civil trial that found him guilty of recklessness, Hazelwood
explained that after the collision, feeling that he'd been "hit in the breadbasket
with a sledge hammer," he "vomited into the commode," a statement duly
reported in the local and national press.[15] In keeping with the tradition of a
sea captain's symbolic union with his ship, Hazelwood's references to his
stomach and to vomiting imaginatively paralleled the rupturing of the tanker's hull. A master navigator of a complex machine carrying oil through sensitive waterways, the chain-smoking[16] Hazelwood not only polluted his own
physiological systems with alcohol but allowed his ship to disgorge a foul and
smelly muck. As freelance reporter Pope Brock puts it, "oil vomiting out of
the wrecked Valdez" transformed "Alaska's Prince William Sound, frisky with
salmon and seals, into a greasy loo."[17]

However, another discourse on Hazelwood appeared in sympathetic bio-
graphical stories in the media, sanitizing to some extent the visceral, excre-
mental images of pollution. Here, Hazelwood became the romantic symbol
of an outmoded laboring masculine body, supplanted by, rather than being
an intimate part of, the alienating complex systems and virtual realities of
huge corporations. Hazelwood was, in fact, an absent body. Instead of being
at the wheel during the time of the collision, he was engaged in paper shuffling
down below. In Brock's article, controversial in Alaska because of its soft treatment of Hazelwood, the former captain told of his animosity against the
"sharp pencil boys," his battle to "maintain some seamanship" amid all the
paperwork, and his refusal to engage in "ass-kissing."[18] Furthermore, twist-
ing the scatological metaphor so that *he* became Exxon's discard, he said that
in firing him, Exxon had "flushed [him] down the toilet in two-a-day press
conferences."[19]

In one of his first interviews, Hazelwood told a group of local and national
reporters that in the spectacle of images, he was detached from his own image and, in viewing himself, felt that he was having an "out-of-body experience."[20] Thus, both media technologies and his abstracted work environment
were implicated in Hazelwood's alienation. He shared with reporters that he
had first gone to sea in a "romantic time," but that now computerized, automated ships have "developed a pretty sterile atmosphere."[21] Maintaining
that it is normal for captains to be below decks, he explained, "It's not the

captain lashed to the wheel or Errol Flynn swinging from the rigging—the common perception. It's a pretty professional operation."[22] Hazelwood was referring to the rapid changes in the shipping industry, reflective of many other industries, in which computerized systems are replacing skilled labor. Sophisticated radar, collision warning systems, and autopilot allow a vast modern tanker to sail by itself with only a handful of crewmembers.[23] In fact, the *Exxon Valdez* was on autopilot when third mate Gregory Cousins realized that it was on a collision course for Bligh Reef, preventing him from steering it. To bring this home, prosecutor Brent Cole compared Hazelwood's use of autopilot to "putting your car on cruise control when you're approaching an accident."[24] The debate over Hazelwood's culpability for the disaster centered on whether he should have been piloting the craft through inland waters, so that his skill might have avoided the collision. And yet trends in the shipping industry are moving away from valuing human expertise, or even human presence.

Thus, Hazelwood emerged in some popular accounts as the tragic representative of a lost way of life, sacrificed to a modern autonomous technosystem. And Hazelwood's story was tied to conceptions of a besieged masculinity, a traditional masculinity associated with certain embodied technological relations. *People* magazine portrayed him not only as a struggling alcoholic, but as a solid suburban citizen and outspoken environmentalist, quoting a "tug handyman" who said that the sea-loving Hazelwood "comes down to the harbor during his vacation, goes out with the lobstermen and helps pull in the nets."[25] Three months after the disaster, *Time* published a very sympathetic article about Hazelwood, suggesting that he was a hero who had actually saved lives by keeping the *Exxon Valdez* stable after the collision. Boyhood friends and classmates spoke of Hazelwood's "feeling for the vessel," his "seaman's eye," and his "sixth sense about seafaring that enables you to smell a storm on the horizon or watch the barometer and figure how to outmaneuver it."[26] And his hard-drinking rugged individualism and refusal to be a company man were blamed for his long-standing problems with Exxon. Hazelwood's "disillusionment" and "powerlessness"[27] in the face of changes to the shipping industry ran alongside his drinking and troubled family relations, particularly his self-acknowledged neglect of his teenage daughter, whom he even discussed in court. An Anchorage journalist concluded that Hazelwood had "worked his way up the chain of command, but by 1985 he was approaching a mid-life crisis."[28] Hazelwood's image was configured in a wider cultural discourse on troubled low-level corporate men in their forties who ignored their families and hated their meaningless, abstracted, alienating work. A massive environmental crisis was thus intimately related to a personal, domestic crisis of masculinity, in transition from a sensual labor to highly organized information work. Coordinator of the Alaska Oil Reform Alliance Mei Mei Evans said of Hazelwood's acquittal on criminal charges, "In some ways, he was just an agent of chance and if it wasn't him, it likely would have been some other poor player. I perceive the real culprit as the system itself."[29]

Exxon officials, deployed from Texas, served as agents of this abstract corporate system with its flows of capital, oil, and information. Discussing the virtues of a modernist, ahistorical, machinic abstraction, O. B. Hardison writes that the Exxon logo was a "concrete poem" that affirmed the link between art and technology, a poem that achieved "the transparency and precision of science."[30] But in communities with oily beaches, the Exxon logo was better known as the sign of the doublecross, and three months after the spill, protesters burned two wooden Xs on a beach outside a resort where Exxon was holding a dinner party. Evoking an obvious symbol of technorational genocide, environmentalists evoked the Nazis, comparing Exxon managers to the "Nazi high command"[31] and suggesting that the oil pipeline terminal had a "Third Reich appearance."[32] Others blamed patriarchy. In her book on feminism and ecology, Joni Seager suggests that the Bhopal and *Exxon Valdez* disasters proved that the corporation's masculine climate of impersonality, rationality, and neutrality led to disaster.[33] However, the apparent technorationality of corporations is a cultural imaginary, as normal accidents such as the oil spill prove time and time again. And the discourse on rationality is a way of restabilizing the corporation's image, even if unintentionally. The image of Exxon that emerged in the wake of the spill was one of a foreign entity, detached from the local environment and its history, an image that would later serve the corporation well.

Unlike Hazelwood and his ship, associated with dirt, vomit, and excrement, Exxon managers were portrayed as the white shirts, afraid of getting dirty or engaging in hands-on labor. In fact, for television appearances, representatives of the State of Alaska wore flannel shirts to set themselves in contrast to the suits and ties. Fighting to get Exxon to fulfill an obligation for the cleanup, they usually appeared on oiled beaches, whereas Exxon execs were interviewed in offices or shown at press conferences. Dennis Kelso, head of the Alaska Department of Environmental Conservation, became a flannel-shirted hero whose "huge how-ya-doin' smile" stood in contrast to the "resentful, sullen, and out of place" demeanor of Exxon execs.[34] When Exxon declared beaches clean, Kelso, or another environmental representative, would dig his hand down in the beach gravel and bring it up dripping with oil. And Alaska officials kept jars of oil, smelling "like the grease pit of a grungy automobile repair shop,"[35] to initiate journalists from other states into the visceral experience of disaster.

Alaskan environmental official Dennis Kelso at the site of the *Exxon Valdez* oil spill. (Photo by the Alaska Department of Environmental Conservation/Allen Prier)

The cleanup itself was dirty, smelly, dangerous, and ultimately demoralizing work.[36] Employees of Veco, subcontracted by Exxon for the cleanup and given a "voluntary" drug test, worked twelve-hour days and had to wear hot rubber suits, hosing down the beaches with sterilizing hot water. Community volunteers who launched their own cleanup began a monumental task of washing rocks one at a time and swirling beach gravel around in plastic buckets. Public sentiment indicted Exxon managers for not physically participating in the work. Satirizing Exxon CEO Lawrence Rawl's infamous apology letter that appeared in about 100 newspapers, a widely distributed Matt Groening cartoon suggested that "Hexxon" arrogantly considered itself too mighty and wealthy to help in the cleanup.

Workers cleaning the beaches with hot water spray. (Photo by the Alaska Department of Environmental Conservation/Patrick Endres)

The class division was also evident between local volunteers' adaptive use of available technologies and Exxon's import of high-tech equipment in a mostly ineffective assault on the oil. The spill occurred on March 24, and five days later, frustrated with a lack of response from Exxon and Alyeska (the oil spill consortium that runs the pipeline), fishermen from Cordova began laying boom to save their fisheries. With the help of the Alaska Department of Environmental Conservation, the fishermen in small boats, called the Mosquito Fleet, "used various creative configurations of absorbent boom, pompoms, and containment boom to build defensive lines whose strength was the sum of many improvised parts."[37] By April 13, oil was unexpectedly moving toward Cook Inlet. Residents of threatened communities there began "constructing booms out of materials at hand: logs, fish-net webbing, typar (a construction fabric) and insulating foam."[38] Complaining of the lack of help from Exxon, a Seldovia fisherman testified, "We started out with home-made tools,

Exxon's high-tech Omni Boom. (Photo by the Alaska Department of Environmental Conservation/Rob Schaefer)

fish totes, barrels, buckets, things we brought from home, home-made dip nets, and by the end of the job, it hadn't improved a bit."[39] Two days later, Exxon's subcontractor, Veco, arrived with money and equipment, too late to prevent the oil from hitting the beaches. The people of Larsen Bay reported that they had developed their own method of oil pickup patterned after seine fishing, a method that was threatened by Veco's appearance on the scene:

> An Exxon representative arrived shortly thereafter with their experts whose job was to train people in deploying boom, safety techniques, and handling oil and oiled wildlife. However, the people they came to train were out cleaning the beaches and corralling the oil in their own makeshift equipment. They had an effective containment system under control, and all of their resource people knew the locations of impacted areas, when to safely go out in boats, and which beaches to protect.[40]

On April 18, having finally arrived in Alaska, Lawrence Rawl declared at a press conference, "We've assembled a very large number of people here, from all over the world, in fact. We've got over 200 vessels, 87 skimmers, 26 aircraft, 250,000 feet of boom in use and another 135,000 available."[41] Although a brief effort was made on the beaches to wash rocks by hand, Exxon mostly promoted its use of high-pressure hoses, chemical dispersants and cleaners, and mechanical rock-washing devices. But residents complained that crucial local knowledge of their waters and shores was being ignored, and this was framed as a battle between drawling pencil-pushing Texans and local, embodied labor. One cleanup volunteer, who had helped develop an innovative geotextile boom, said, "They [Exxon] act as if using your hands and your body to do work is primitive. They think that since we are people who make our living in a physical way, we must be dumb."[42] Another complained that Exxon dismissed the intelligence of innovative locals: "I don't think Exxon ever gave them the credit for being more than a bunch of fishermen coming in with oil underneath their fingernails."[43] Thus, the *Exxon Valdez* spectacle played out its familiar class lines between those who were willing to get dirty and those who weren't.

Good Housekeeping

None of the various "skirmishes" in the "guerrilla war"[44] against the oil really worked, and neither containment nor cleanup efforts prevented lasting deposits of oil and long-term damage to Alaska's wildlife populations. The debate devolved into arguments over the word *clean*. In an early press conference, Rawl declared, "We, in fact, will have most of this—these beaches and these rocks cleaned by mid-September, 1989."[45] When it became obvious that the beaches could not be returned to prespill conditions, Exxon officials began referring to "treated," "environmentally stable," and "environmentally clean" beaches. For this, Exxon was awarded the sixteenth annual Doublespeak Award from the National Council of Teachers of English, citing the company's argument that "clean doesn't mean every oil stain is off every rock. It means that the natural inhabitants can live there without harm."[46] Once again evoking the contrast between household technologies and Exxon's cleanup efforts, a Kodiak resident pointed out, "A lot of these beaches may appear relatively clean on the outside, but if you use a garden shovel, a common garden shovel and just dig down a foot, they're just laced with oil and mousse."[47] In their public relations brochures, Exxon implied that people were misinterpreting visual evidence such as a photograph of Eleanor Island: "The dark coating on the rocks in the foreground is primarily black lichen."[48]

In the national media, arguments over "how clean is clean," and contrasts between high and low technologies took on a gendered imagery. Curiously, the cleanup became symbolically associated with women's work, at which male Exxon managers were inept. One participant at a protest against Exxon in Seattle said, "I got kids, and I try to teach them to clean up their own mess, you know? What I want to know is, how do I make them believe that when nobody makes Exxon clean up their shit? It's like telling your kids, well, there's rules unless you're rich."[49] In a Tom Toles cartoon an Exxon executive, wearing an apron over a suit and tie, appeared washing a rock in a plastic basin, putting the rock on an oil-soaked beach, and declaring the beach clean. Another local cartoon in the *Anchorage Daily News* showed a babushka-clad woman cursing and scrubbing rocks as "Mother Nature, Exxon's Cleaning Woman." An essay in *Ms.* blamed Exxon managers for living in a male-dominated system where wives and servants always cleaned up after them, and suggested that they be given a "basic housekeeping tutorial" and spend a year in Alaska "with brooms, mops, and squeegees. And paper towels to clean the birds."[50]

Housekeeping was associated with the feminized space of nature. Taking on some very traditional metaphors of the American wilderness, the media

MOTHER NATURE, EXXON'S CLEANING WOMAN.

1989 cartoon from the *Anchorage Daily News* depicting gendered images of the spill. (Cartoon by Peter Dunlap-Shohl/*Anchorage Daily News*. Courtesy of the *Anchorage Daily News*)

routinely compared Alaska to a beautiful woman, commenting on the extent of Big Oil's abuse of "her." In the national media, Alaska took the form of pristine virginal wilderness defiled by Exxon, a whore who had sold herself to Big Oil, a middle-aged woman who now had a wrinkle, or a predisaster Mother Earth in a "procreative fury"[51] of whelping, spawning, calving, pupping, and hatching. Alaskans talked of Exxon's rape and despoilment of their waters, and held protest signs that declared, "Sterilize Exxon, Not the Sound." An Alaska tourist organization, trying to revise this image of a damaged beauty, developed an infamous ad showing Marilyn Monroe as Alaska, with a "beauty mark" representing the spill.[52] Referring to this image, Valdez mayor Lynn Chrystal told a community booster organization, "I would prefer to think of Prince William Sound as a beautiful woman with a black eye; very sad and ugly to look at, but it will heal and fade away with time, and Prince William Sound will once again be a beauty to behold."[53] Chrystal's statement was appallingly insensitive considering the rise in the numbers and severity of domestic violence cases reported by crisis counselors in Valdez and other oiled communities.[54] The conditions for abused women and children, who suffered an increase in broken bones during the stressful cleanup, were not much covered in the media, especially the national media, and their experience can be gleaned only through number-crunching sociological reports funded for use in litigation. Although these bodies represent the human cost of the *Exxon Valdez* oil spill, their voices do not emerge.

Like the construction of Hazelwood, cultural negotiations of gender and technology unfolded in the bird and otter rescue centers. In both cases, the disaster provided an arena for restabilizing traditional gender/technology relations. The Exxon-funded rescue centers were feminized social worlds, staffed mostly by women, where domestic relations were constructed and maintained through daily manual labors and simple artifacts. In fact, women were specifically solicited for the jobs. As one local newspaper announced, "Alice Berkner [a Berkeley ornithologist brought to Alaska for the cleanup] is looking for a few good women—and men—to help clean oil-tainted birds."[55] Taking over classrooms and gymnasiums of local schools, the rescue centers represented domestic spaces where intimate relations between humans and anthropomorphized animals were established and preserved. The wildlife rescue teams used household technologies. Birds and otters were "laundered," placed in warm soapy water and scrubbed: "Workers lathered them up with detergent and water, working the suds into the fur with their fingers."[56] In contrast to the abrasive, sterilizing hoses, these feminized technologies evoked a pleasurably sensual contact: "Nancy . . . rubbed the sick otter pup with terry cloth towels, trying to imitate the slow, soothing strokes of its lost mother."[57] Rescue work-

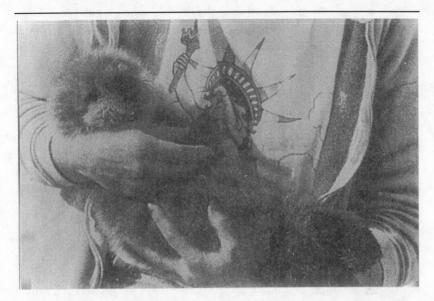

A worker from the Valdez Otter Center holding a cleaned animal. (Photo by the Alaska Department of Environmental Conservation/John Hyde)

ers used Water Piks, Dawn detergent, rubber bands, toothbrushes, wooden dowels, plastic buckets, sheets, cotton swabs, trash bags, shot glasses, plastic gloves, tweezers, and cardboard boxes. Rescue volunteers sentimentalized their work with otters as maternal: "Being a mom, I wanted to make each of these little hurt beings feel better. You know, they are a lot like kids."[58]

Some women scientists tried to dispel the image of rescue workers as "little old ladies whose kids left them with an overabundance of maternal instinct,"[59] and replace it with a discourse on "bird husbandry, ornithology, veterinary medicine, and surfactant chemistry."[60] But, as at Chernobyl, these scientists were deployed to mediate the disaster, to make it productive, and because of the deployments of gender in the imagination of this disaster, to bridge the lab and the household. Many highly visible women scientists were associated with the spill cleanup, including Jessica Porter, chief veterinarian of the Valdez bird rehabilitation center; ecologist Page Spencer, injury assessment coordinator for the Kenai Fjords National Park; biologist Lisa Rotterman, who tagged and counted otters; oceanographer Sylvia Earle, who conducted a spill assessment for the National Oceanographic and Atmospheric Administration (NOAA); and Frederika (Riki) Ott, an expert in marine sediment pollution.

Earle was the most famous of these, having made her popular reputation in 1970 as leader of an all-woman scientific team that stayed in the Tektite II undersea habitat, used for oceanographic research and isolation studies.[61] She later went on to break the record for the deepest dive in a Jim suit, and remains an advocate and developer of high-tech diving equipment, including her one-person submersible for doing "barrel rolls with the whales."[62] In her account of her visit to Prince William Sound after the spill, Earle described a fecund nature, in the midst of her "outrage" provoked by the "shrill, pained cries of sick and disoriented victims."[63] Flying over the sound she saw "tens of thousands of slim, silver-blue male fish [jetting] milky life-giving clouds of sperm over acres of eggs plastered by their female counterparts on rocks, shells, and seaweed," and rocks "embroidered with what appeared to be millions of perfect seed pearls," or herring eggs.[64] Earle's message was the resilience of nature in the face of catastrophe, a message she would echo in her trip to the oily waters of the war-ravaged Persian Gulf. Walking along oiled Alaska beaches with a rubber-clad Earle, a journalist for the *New Yorker* wrote, "She bent to examine amphipods, copepods, whelks, and other small creatures, and at one point she reached into a small pool and picked up a tiny starfish on her fingertip—an apparent casualty that soon began to show signs of life."[65] Alaskans distrusted Earle as a representative of the NOAA (she would later become head of the organization), but the national media was fond of Earle for her tales of the sea that have an environmental ethic, yet present technology as a means of access to a sensual environment. Thus, she was a fitting national actor to play the healing Gaia whose touch brought the world back to life, despite the ravages of corporate and military environmental disasters.

Riki Ott was a local hero, known for predicting a spill at a community meeting only hours before the *Exxon Valdez* landed on Bligh Reef. A fisherman and artist as well a scientist, Ott remained a vocal presence after the spill, often quoted as an expert in the national media, critiquing Exxon's scientific methods in its spill assessments. In a report for Greenpeace, she accused Exxon of begging the question, beginning with a "systemic bias wherein a large and substantial sampling effort was expended, but the effort seems to be intentionally directed at 'proving' there was no effect from the spill."[66] Ott also quilted a wall hanging commemorating Prince William Sound, complete with fish, birds, puffins, and an otter female with a pup on her stomach. "Slice of Life" became so well known that it is still sold as a postcard in Valdez. In a *National Wildlife* spread, Ott appeared with a pile of books on toxicology, in front of her quilt, in a liminal position between these two worlds of the domestic arts and scientific research.[67] Ott's image served as a bridge between gendered knowledges, techniques, and practices. Women scientists thus made it possible to restore an image of a reproductive, healing science against the machinations of Big Oil.

In the rescue centers, biology and home economics were leveled at injuries to wildlife, imaged as a kind of domestic violence. *Tikkun* writer Laurie Stone, volunteering in a Valdez rescue center, wrote that she was galvanized by the "innocent, helpless otters" and a "dead baby seal with blood pouring out of its mouth."[68] These she compared to Hedda Nussbaum's child, who had been beaten to death by Nussbaum's boyfriend, a case that was getting media attention at the time. The first otter pup rescued from the spill was named after Little Orphan Annie. Many descriptions of the rescue attempts offered images of young or female animals as victims of a sadistic corporate power that destroyed harmonious family structures. Rescuers viewed themselves as "family" while monitoring the bodies of otters and birds, and saving them from Exxon's disastrous science and its "political and bureaucratic machinations."[69] The centers thus became sites of atonement for masculinist corporate sins against women, children, and nature. Writing in *Rolling Stone*, Tom Horton described his experience in a rescue shelter, where he saw an oil-blinded female otter, which had aborted a pup, being licked and apparently comforted by another otter. Horton concluded that Exxon executives could atone for their sins against the environment by serving as "volunteer otter handlers."[70]

The anthropomorphized otter was marked with the special maladies of a maximalist technological culture, even afflicted with the same vices such as gluttony and overpopulation: "The otter itself is a keystone predator, a species rather like us in that its voracious appetite for shellfish is capable of causing precipitous change in a habitat. Left alone, it will increase its numbers beyond the capacity of a habitat to support it."[71] The oiled otters suffered from physiological damage that sounded familiar: shock, anorexia, emphysema. Subjected to the pollution of an oil-dependent society, the otters clawed themselves frantically and self-destructively, and, because of brain lesions, moved aimlessly, bumping into things. Their livers were destroyed in their grooming efforts. They sometimes became violent, even attempting to drown their own pups. If returned to an unfamiliar part of the sea, the otters suffered a high mortality rate. And they carried a herpes virus, spread in the treatment centers, that afflicted their mouths and genitals. Sexually transmitted diseases, weakened immune systems, domestic violence, psychological disorientation, chemical toxicities: these were all reminiscent of a larger cultural discourse on the dangers of a crowding modern technology.

Whereas the bodies of human disaster victims are usually not represented in disaster coverage, the bodies of animals could be depicted graphically:

"One otter, intensely ill, was not grooming—in fact not moving at all. He lay propped against the wire mesh of his pen, shuddering, staring."[72] Wildlife experts visiting Prince William Sound described walking through oily water, treading on unidentifiable bodies, "the grisly masses of oil and feathers and fur that had once been animals."[73] They described otters with their paws bitten off and eyes scratched out, oil trailing from their anuses: "We saw two cases of severe trauma—chewing the skin right off their feet right down to the bone and rubbing the skin right off their ears."[74] John Keeble, critiquing the photographing of oiled otters as "voyeuristic," went on to describe walking among oiled rocks, his feet touching "something too hard to be liquid and too soft to be rock, a pliant thing down there, too solid to be wood or stone, a thing that still had the give of what had been a life and the integrity of lineament, skin, and bone, a thing that slithered away from the boot and when punctured gave off gas."[75] The oil spill provided an analytical site for exploring the carnage of technological disaster through the bodies of animals and birds, which are not veiled with the same cultural sanctions against nonfictional representation of violent death as humans are.

The rescue centers, as domestic realms, sought to heal these terrors through mothering and to provide good housekeeping for the smelly mess rupturing from the tanker. The middle-class household serves as an imaginary retreat from the stress of public life. During the cold war, it was reconfigured as a fallout shelter where the nuclear family could be saved from disaster and continue its reproduction, its baby boom. Since the 1960s, environmentalists have also used the image of nature's household as a refuge from or corrective to technological disaster. Feminist peace activists and their philosophical offspring, ecofeminists, have developed a protest iconography of cooking, weaving, quilting, gardening, bread-baking, and other domestic activities that would serve to offset a "male culture making"[76] devoted to competition and war. Carrying quilted and appliquéd banners and weaving webs of yarn into fences around nuclear repositories was a favorite protest method in the 1980s. This view of household activities carries the traditional Western idea that women are closer to nature and because of biology or social influence, more caring and nurturing and peaceful than men. Over the past few decades, in many currents of discourse, the household emerged as the feminine space of refuge from complex masculinist technological systems and the traumatic stress that accompanies them. (On the other hand, in the 1990s, the household is being reconfigured as a site of extreme technological violence involving ordinary artifacts, such as Lorena Bobbitt's knife and O. J. Simpson's bloody glove.)

The household safe from technological risks is entirely imaginary, without basis in the material conditions of women's lives. In the context of the *Exxon Valdez* spill, the evocation of housework has ironic implications. People working in the household do indeed have complex interactions with tools, machines, and complex systems such as plumbing and electricity that shape experience and consciousness. As technology historian Ruth Schwartz Cowan points out, modern housework has become increasingly time-consuming and demanding, especially with the advent of the automobile.[77] Housework now depends on the gas-guzzling car for shopping and delivering children. And household chemicals are often toxic and petroleum-based. Rubber bands and plastic buckets are made from petroleum products. During the spill clean-up, Dawn detergent, promoted by workers as a little household secret for cleaning birds, was actually being vigorously advertised by Procter & Gamble's PR firm, Manning, Selvage & Lee, who shipped it to the rescue sites and made promotional videos of bird scrubbing. Thus, rather than unfolding in a morally pure context, housework fully supports and participates in oily networks of pollution and consumption. And the domestic space is by no means free of appetites and violent desires.

While the rescue centers' work with otters stood as a maternal injunction against modern technological interventions, the discourse recognized these ambiguities. Work at rescue centers was perceived not only as therapy for otters, birds, and local people suffering from posttraumatic stress disorder, but atonement for consumers from the rest of the country. Interviewed in his home town after his work in an otter rescue center, one San Diego veterinarian explained, "When you really start to think about it, since we all are using oil products, we all have to assume some responsibility."[78] Another volunteer, pregnant with her first child, said, "I drive a car; I use oil. . . . In a sense, I share in the blame for this. I had to do something."[79] There was great concern that the otters and birds, rather than being completely saved by these efforts, would be tainted by the technologies of rescue and would carry away with them the gluttonous sins of technological culture. Furthermore, there was no wild untainted space for them, and many otters that were returned to the ocean died.[80] Others were shipped to zoos and aquariums. The terrible bargain of survival was that survivors would remain "under the influence," dependent on artificial worlds where they were provided with grooming and "fed rich diets of crab and fish."[81] Thus, the household was implicated as a locus of consumption and dependency and, like the lifeboat, could provide no safe shelter from technological risk.

Shangri-La

Whereas otters provided an analytical site for examining the physiological and psychological burdens of living with high-risk technologies, another actor appeared as a representative of the pristine wilderness, the Alaska Native villager who lived in a "Shangri-La." The oil spill was often said to have ruined Shangri-La, that green valley "safely far away from the problems of the twentieth century."[82] In James Hilton's novel *Lost Horizon,* adapted to film by Frank Capra, Shangri-La, an agrarian utopia, lies in a Himalayan valley, founded by a Jesuit priest wishing to escape the perils of modern war, including the atomic bomb. Living in a lamasery on a hill, the holy men import the art and literature of Western culture while ruling the indigenous people below through principles of moderation and curtailing of appetites such as hunger and lust. The comparisons of Alaska to Shangri-La evoked not only a sheltering of refugees in an Edenic wilderness safe from the disastrous machine, but the rational management of Alaska Natives.

The *Exxon Valdez* disaster offered a new version of the fall from the Garden. Because maximalist technological cultures perceive indigenous people as nontechnological, they appear to survive in a pure, transparent relationship to nature, in the Garden. With little understanding of the modes of production or networks of exchange in a subsistence economy, or recognition of technology transfers and adaptations in these villages, the media and some social scientists preserved a vision of innocents overrun by a completely alien technological and economic system. Native communities have economies that blend wages and subsistence, are not necessarily conservationist in their goals, use fossil fuels, and have a historic relationship with Big Oil that includes negotiations over land claims. But representations of Alaska Natives as being overrun not only by oil, but also by modern "civilization" and its corrupting technologies and values, were driven by the spectacle's terms of alienating work, consumption, and addiction. Threatened by pollution from the spill, subsistence ways of life practiced by Native communities became a benchmark for judging the corrupting influences of money and oil.

The spill affected some 1,000 Alutiiqs, who depend primarily on fishing, hunting, and gathering shellfish for their daily existence, social networks, and spiritual well-being. The "Native perspective" on the spill most often came from a speech by former Alutiiq chief Walter Meganack, a longtime political activist who had helped negotiate the Alaska Native Claims Settlement Act and had helped bring electricity, plumbing, sewage, and better housing to his community. Describing the celebrations usually surrounding the annual spring catch, Meganack said of the spill's disruptions, "It is too shock-ing to understand. Never in the millennium of our tradition have we thought it possible for the water to die, but it's true."[83] He described a community "invaded by oil companies offering jobs, high pay, lots of money," and the replacement of communal harvesting activities with cleanup employment under a hierarchical command devoted to "busywork."[84] The speech ends on a note of hope, describing the community's many battles with disease, alcohol, and drug abuse and its ability to survive both these and an even more devastating incursion of destructive forces. Often printed whole or in part in local publications on the spill, the speech became a rallying cry for spill communities, both Native and non-Native, disgusted at Exxon for its pollu-tion of their life-sustaining waters and its ineffectual cleanup.

But Meganack's words were also subsumed into the national spectacle, where Alaska Natives were constructed as symbolic victims like the otters,[85] overrun by a faceless capitalism disgorging not only oil, but cleanup money and processed foods that entered susceptible communities like addictive drugs. Journalists emphasized a rise in alcoholism and portrayed Natives as suffering from a moral failing in their participation in the cleanup. First sug-gesting that Native communities were "faced with extinction," *Newsday* rep-resented Chenega villagers as eating Spam and drinking soda pop (that they admittedly didn't like) while sitting around waiting for orders, whittling, and throwing rocks: "The subsistence fishermen have become shrewd business-men. Instead of working with fishing nets and hunting rifles, some walk around the dock with radio transmitters and discuss logistics of boom-lay-ing operations. Many are paid $16.69 an hour by Exxon to wait at home as 'standby workers.'"[86] In the *New York Times,* Timothy Egan, reporting on the tiny community of English Bay (now called Nanwalek), suggested that the Exxon cleanup had rapidly introduced modern technologies that were cor-rupting the people there: "Instead of hanging salmon to dry this month, as Aleut Natives have done for centuries here on the toe of the Kenai Peninsu-la, John Kvasnikoff was putting up a $3,000 television satellite dish on the bluff next to his home above the sea. A few houses away, Jeff Evans, his hair painted green, churned up the road with his new $1,500 'Fat Cat' motorbike, portable stereo headphones over his ears. In years past, he would be storing food for the winter."[87] This set of binary oppositions between traditional practices and modern technologies oversimplified the history and dynam-ics of technology transfer in these communities, the active selection and embedding of technologies within them. Perhaps in its own fear of televi-sion's ascendance over print media, the article emphasized the purchase of satellite dishes and the subsequent transformation of hardworking people into "couch potatoes." Television and Exxon, in an apparent collaboration, destroyed traditional family values: "The 21-year-old daughter of the chief who was wearing an MTV T-shirt . . . said she left behind the chores of caring for her infant daughter to work for Exxon."[88] Furthermore, Egan claimed, stress over the spill was causing a rise in alcoholism that he associated once again with the corruption of money and media technologies: "[One resident] earned $22,000 from the Exxon cleanup and is thinking of moving to Anchorage. On a day when he would normally be hanging fish in the smokehouse, he was playing the Sex Pistols, a punk rock 'n' roll band, at high volume. Sober for nine months, he is drinking again."[89] Not surprisingly, the residents of English Bay were unhappy with their portrayal in the press, which they viewed as negative and inaccurate.[90]

The conflation of the oil spill with MTV, satellite dishes, motorcycles, wages, drinking, and the Sex Pistols suggests a larger narrative agenda than the realities of life in English Bay. Here and in other early accounts of spill, the emphasis on alcoholism, rather than on many other environmental and social aspects of the spill, configured Alaska Natives in a similar story as Hazelwood's. Addiction and dependency, erosions of family structure, and loss of embodied knowledge resulted from a Faustian bargain with Exxon, the sign of the modern technosphere. This set of conventions stems from an apocalyptic Western view of industrialism and industrial disasters that since the nineteenth century have been accompanied by stories of social disintegration, violations of social mores, and contagious feelings of isolation and withdrawal.[91] In his neo-Luddite defense of the Unabomber, who killed a Burson-Marsteller executive in anger over the *Exxon Valdez* spill,[92] Kirkpatrick Sale writes, "Whatever material benefits industrialism may introduce, the familiar evils—incoherent metropolises, spreading slums, crime and prostitution, inflation, corruption, pollution, cancer and heart disease, stress, anomie, alcoholism—almost always follow. And the consequences may be quite profound indeed as the industrial ethos supplants the customs and habits of the past."[93] He goes on to discuss the Ladahkis of northern India, who supposedly no longer sing traditional songs because of the introduction of the transistor radio.

Because the narrative of technological progress is linear, indigenous peoples are frozen in some original moment of naked, face-to-face existence in which experience is not technologically mediated. Through this construction, people in maximalist technological cultures set up an imaginary compass point for assessing their own position in the narrative of progress, and their own anxieties about the inherent dangers and losses in interfacing with their assemblage of machines and systems. Technological disasters are figured as a rupture in the assemblage, allowing for a glimpse of a Garden to which the citizen of the technosphere might return, sloughing off the apparatus with all of its pleasures and horrors. Of course, there is no Garden to which anyone can return because, as Don Ihde points out, all human activity is technologically embedded.[94] He suggests, following Heidegger, that there are different modes of revealing the world, different ways in which cultures embed technologies in a dynamic exchange. The transistor radio may be a cultural instrument, bringing with it a set of human-technology relations, but its reception is not predetermined. Nor is the reception of an oil spill predetermined by the culture of the spillers.

Joseph Jorgensen makes a similar argument about the social science surrounding the *Exxon Valdez* damage assessments, in which Native culture was reified into a thing frozen in the past. In his stern critique of the studies used by lawyers on both sides of the case, Jorgensen argues that social scientists used sloppy methods and faulty arguments about Native culture. Because they presented Native culture as a "thing that can be damaged," social scientists did not bother to analyze the specific forms of severe harm done to dynamic social practices, as Jorgensen did in his much more sensitive and complex study of Native communities for the Department of the Interior.[95] Jorgensen's own work suggests fundamental ideological differences between a Western capitalist system and the "traditional-communitarian" societies of Alaska Natives, stressing their adaptability to new technologies. Because other social scientists reified the concept of culture, the court could dismiss the claim that the spill had destroyed or damaged a whole way of life.[96] Jorgensen argues that the spill was not pollution or damage to a culture, but to the environment, and that the culture responded from its own understanding. Jorgensen is particularly hard on Exxon consultant and anthropologist Paul Bohannan, who argued that Natives were an American ethnic group, rather than a culture, and that subsistence was not a way of life, but a benchmark used by Natives to set themselves off from other Americans. Native culture, Bohannan suggested, was a thing that had been "smashed" much earlier by colonization. Therefore, Natives could not claim damage to a way of life, just as other groups could not. As Jorgensen points out, Bohannan, by his own admission, based his conclusions on scholarly literature, newspapers and magazines, and government and organizational studies, and never talked to a single Native villager. Because lawyers for the villagers would not allow any but their own social scientists to visit, this was not Bohannan's fault. But it does reveal why Bohannan would draw the conclusions he did because in the logic of the spectacle, subsistence *was* a benchmark, not for Natives, but for outside observers, who configured Natives as a state of cultural purity by which technological pollution could be measured. The *image* of subsistence, rather than its actual social practices, was necessary to the story of corruption played out in the larger narrative of disaster.

The same is true of damage assessors' use of posttraumatic stress disorder (PTSD), an easy descriptor for the complex effects of the spill on highly diverse communities. A study conducted by Impact Assessment, headed by medical anthropologist Lawrence Palinkas, found that inhabitants of oil-spilled communities were suffering from PTSD, especially young people, women, and Natives. The study was flawed for several reasons. For one thing, using the diagnostic tools for PTSD already begged the question of whether the oil spill was a "traumatic event" that was "outside the range of usual human experience," as required by the American Psychiatric Association's 1987 definition of PTSD. As anthropologist Allan Young explains in a critique of PTSD, diagnosticians' identification of a traumatic event is highly problematic because they have a culture-bound idea of what is "usual": "One would have to know something about the meaning of the event in context: the frequency with which it occurs or occurred in the person's community; the community's moral order, that is, its ideas about what is good, tolerable, forbidden, and salient; the degree to which the person understands and identifies himself emotionally with this moral order; his level of self-awareness, empathetic awareness, and moral autonomy; and so on."[97] Rather than exploring in depth whether and why the spill was traumatic for different persons and their communities, and allowing these voices to emerge in full, damage assessors for the Alaska Conference of Mayors laid the same grid on all communities, leading to the conclusion that there was a "99% increase in post-spill PTSD" in "spill-affected communities" as compared to "non-affected communities."[98] Whether there could possibly be purely "non-affected communities" also begs the question. Although the report briefly discusses changes in subsistence practices, these differences are readily subsumed into meaningless numbers. For example, if the stressor was the oil spill, how could there be a 99 percent *increase* in PTSD? From what state of events could an increase be measured? To what degree were inhabitants of Prince William Sound already living in an increasingly polluted, stressful, and environmentally damaged world—not a pristine state but one in which oil leaks and other such events were already sadly normal? The decontextualization of the spill and its definition as "outside the range of human experience" set aside more complex social questions about compromise, resistance, and acceptance in bargains with Big Oil already in place. The *Exxon Valdez* oil spill was an acute manifestation of the results of these bargains, a worst-case scenario that threw them into sharp relief.

Second, the stories that might have emerged were suppressed in what Young calls the "harmony of illusions." PTSD is made real only within psychiatric techniques, and *then* penetrates people's lives. For example, in the aftermath of the spill, some psychologists on "Critical Incidents Debriefing Teams" were telling people they had PTSD while others were seeking to determine through questionnaires whether the syndrome was emerging organically in people's feelings, behaviors, and dreams.[99] These objectives were clearly at odds. The discourse on PTSD folded back into people's understanding and self-knowledge, as when a Port Graham resident told a reporter, "It was an unbelievable incident that happened. I still suffer from post-traumatic stress syndrome from the *Exxon Valdez*."[100] Almost everyone associated with the spill was claiming PTSD, including a former Exxon ship's officer who said that Hazelwood's abuse of him had caused his PTSD[101] and the *Exxon Valdez*'s second mate, who said he had retired after the spill because he suffered from PTSD.[102] The Oiled Mayor's damage assessment was conducted almost an entire year after the spill, and certainly many of its subjects already knew of PTSD and its symptoms and could construct and reduce their own experience within this external, rudimentary template.

Third, although Impact Assessment used other indicators in their report to identify changes in Native subsistence practices, such as hunting, fishing, gathering, and sharing of food, and to distinguish them from non-Native social practices, the "depression" and "anxiety" of Natives were figured as being the same as those of non-Natives. Natives were said to be suffering *more*, rather than qualitatively *different*, responses: "Natives had higher rates of depression and anxiety than non-Natives." This conclusion created a trajectory of harm, from least harmed to most, rather than establishing important differences between *interpretations* of harm, based on Native voices. The absence of Native voices, other than as sound bites, is striking in the voluminous studies and reports surrounding the disaster. And subsuming these voices into statistical analyses, even if done with the best intention of supporting legal claims, does little to increase understanding or sympathy for the violence done to these communities. That Natives were reluctant to talk to outsiders about their reactions to the spill is not surprising in light of the media coverage of their communities. As Jorgensen argues, differences between Natives and non-Natives are significant in "every idea, every ethic, every sentiment, every activity," and the organization of these differences matters, rather than degree.[103] But Impact Assessment's position that depression over the spill was generic, a matter of statistically proven degree, allowed it to give its own entirely institutional (and self-interested) recommendations for Native communities. First on the list: Increase mental health services.

Later, this report was summarized in the journal *Human Organization* and
its ideological assumptions became more apparent. Comparing their subjects
to survivors of the Mount St. Helens eruption, the Three Mile Island accident, dioxin contamination at Times Beach, the Vietnam War, and even nuclear war, the study concluded that Native communities were experiencing
rapid and even permanent social disintegration. Native communities were
called "'pre-contact' sociological systems"[104] so that their difference from
these other survivors was implicitly maintained as the baseline for revealing
the invasive, destructive consequences of "the larger Euro-American social
system." This idea that Natives were somehow "precontact" was absurd,[105]
compelled by the logic of the disaster spectacle that reenacted colonization
for the nervous, stressed-out inhabitants of that urban "technological sophistication." This cultural bias is also apparent in Dyer, Gill, and Picou's study
of Cordova, in which they call the oil spill cleanup a "cultural pollution" of
Native villagers because of the influx of cleanup workers and the introduction of "nontraditional foods and plastics," the imposition of a "technologically sophisticated normative system, which is typical of the urban environments of most American cities."[106] The worst aspect of these reports is that
they present Native peoples as passive, helpless recipients of new technologies rather than active negotiators and builders.

In their own reports on social impacts, Native communities discuss the
emotional effects of finding their beaches, their basic food sources, under a
sheen of oil: "Some who walked the beach [at Akhiok] said it was like a dream
to see oiled birds and feathers lying about. Emotionally, the spill was hard
on these people who depended on the sea for their livelihood. The death of
wildlife was especially hard on the elders. One elder said she cried when picking up dead birds, and wondered 'What's coming of our world?'"[107] But these
reports also emphasize people's active roles in coping with the disaster and
interpreting its consequences, rather than presenting passive victims invaded by technological forces. Villagers deployed their own technologies, skills,
and knowledges in coping with disaster. In Port Graham, women went out
in skiffs and collected chitons before the oil hit, set up an assembly line to
clean and cook them, and distributed them to families. Told that they were
low priority in the cleanup, Chenega Bay residents contacted "everybody in
Government they could think of"[108] until their village became high priority.
Furthermore, they insisted that they be represented on cleanup monitoring
teams. The residents of Akhiok began sending in their own reports to the local
radio station after noticing inaccuracies in its reportage of spill effects. Communities noted Exxon's and Veco's racist practices in leaving the cleanup of
their communities until last, ignoring their knowledge of local conditions,

and making racist remarks about Native workers. At Larsen Bay, workers were offered $10 an hour instead of the usual $16.69: "The community reacted in defiance, and learned quickly to stand up for their rights in further spill response efforts. They revealed their strong convictions in wanting to cleanup the environment which supported their life whether or not they received compensation."[109] As mentioned earlier, the inhabitants of Larsen Bay had developed their own methods of oil cleanup based on seine fishing technologies: "Bailers and five gallon buckets with holes in the bottom were used to scoop up the oil for transfer to the hold of a salmon seiner. More than 700 gallons were picked up in five hours, and 6,000 gallons were scooped up during the first week. The volunteer crew collected an average of 20–30 drums of oil a day."[110] While similar efforts by Cordova's commercial fishermen became a national media event, the Larsen Bay cleanup passed by in silence.

Differences in cultural perspectives were not so clearly drawn along the traditionally accepted divides between nature and culture, bodies and technologies, rural and urban, myth and science, Native and non-Native. In his controversial critique of the limited and mostly inconclusive environmental studies following the spill, Jeff Wheelwright discusses the response in Native villages to the 1989 recommendations of the Oil Spill Health Task Force. A former science editor for *Life*, Wheelwright intended to prove that science had been corrupted because it had been appropriated by litigious interests and that the spill had done remarkably little damage to the environment. Therefore, he was accused of siding with Exxon, especially after he wrote a letter to the editor of the *New York Times* given the headline, "Exxon Was Right. Alas."[111] The problem with Wheelwright's book is not so much his often convincing assessments of bad science, but his subsuming of supposedly good science into a mystical narrative agenda in which he suggests that the oil was not only ultimately benign, but also actually stimulating to the environment. Thus, in his introduction, he describes a dive into the hull of the *Exxon Valdez*, in which "awed" biologists discovered many varieties of sea life, a "cascade of biological enrichment due to the oil."[112] The role of popular science in mitigating the disaster with pronouncements of its life-giving properties, supposedly "proven" with scientific expertise, has become a convention. Like the popular accounts of the Chernobyl biologists, Wheelwright's book fulfills the new narrative expectation that scientists will find a Gaia in the middle of Hades.

Overstating his case in a similar fashion on many occasions, Wheelwright compares himself to the Chugach people, who "gathered . . . facts empirically," through observation of what had gone before. But what he means by empiricism is really a sensuality that he thought uniquely marked a

Native way of life. This is apparent in his discussion of the organoleptic tests recommended by the Oil Spill Health Task Force for assessing the edibility of seafood. The task force told village residents that if the seafood in question didn't look or smell oily, it was safe to eat. Some Native villagers, who insisted that the state do more testing to see whether the oil was getting into the food chain and whether undetectable levels of hydrocarbons might be carcinogenic, interpreted this as racist. According to Ernest Piper, the State of Alaska's liaison to oiled communities, "What they wanted, of course, is what every American living near an industrial site wants to know: Am I safe? Are my kids safe?"[113] After further tests by the FDA and Exxon, the NOAA declared that subsistence foods were safe if they passed the organoleptic criteria.

However, the task force compared levels of carcinogenic polynuclear aromatic hydrocarbons (PAHs) in tainted salmon with those in cured, smoked fish, and declared that the latter had PAH levels 10,000 times that of the former. In what will now seem a familiar postdisaster expert strategy, the task force attempted to frame the spill impact in terms of the technological dangers of everyday life. According to Wheelwright, Alaska Natives "resented" the notion that their techniques for preserving fish were somehow more dangerous than the oil spill, and this he dismisses as a somewhat misguided comparison between voluntary and involuntary risks.[114] But it was more than perception of personal risk, or even damage to a "lifestyle," that influenced a rejection of this argument. The debate was unfolding in a much larger sphere of cultural negotiations in which these tests were never politically neutral, either in original conception or in dissemination of the results. In Piper's account, villagers were understandably angry at the presence of Exxon scientists in the testing. In the political milieu, if people had accepted the argument that their smoked salmon was more dangerous than tainted salmon out in the oiled waters, they would have tacitly accepted an Exxon-influenced philosophy that normalized big technological risks. As Exxon's board of directors put it in their 1989 letter to their shareholders, "Because of human factors and the technical complexity of what we do, it is not possible to eliminate operating risks altogether and still continue to find, produce and deliver the energy resources essential for economic activity and well-being."[115] Referring to "flowerpot" environmentalists and a judge who compared the *Exxon Valdez* disaster to Hiroshima, the industry publication *World Oil* argued, "Only a fool or a New York judge can believe we can get along without the stuff that fuels the world, or that oil industry operations can be made 100 percent risk free. Even supersafe NASA couldn't attain Utopia."[116] As the *Exxon Valdez* polluted some 1,300 miles of coastline, with a heavy toll on wildlife, smoking salmon was certainly insignificant compared to these normal "operating risks" and acceptable losses.

For one reason or another, Wheelwright claims, people guardedly went back to subsistence harvesting and fishing, and when he visited Chenega Bay in 1991, he felt a "flash of pleasure . . . that some people were relying on their senses at last." Wheelwright theorizes, "It is ironic and telling of their situation in the world that Natives turned away from proofs that depended on the acute application of their senses; but this was because they and their advisers had gathered a little bit of knowledge about the sublethal effects of oil on organisms."[117] But in what way was this "ironic"? Only in a worldview in which Natives took up a role as sensualists who could not fully use or assess the complexities of an abstract science. As Wheelwright implies that a little scientific knowledge was a dangerous thing, he polices certain modes of inquiry. Because his overall argument is that an infinitely resilient nature can be left to take care of itself, and because Natives are configured as a more intimate part of that nature, they should also have been left to take care of themselves without the interventions of science.

This rigid view of culture denies the necessary reordering of the world in disaster's wake. As Kai Erikson writes in his study of traumatized communities, the inhabitants experience radical transformations of identity, social relations, and worldviews.[118] The assimilation of disaster is thus a dynamic process of renegotiation with and adaptation to the new terms of existence. Jorgensen suggests that the spill evoked a deeper consideration among Native villagers of the meaning of their environment and led to a wider network of distribution and visiting.[119] Disagreements over that meaning have emerged in the Trustee Council offers of Exxon settlement money to Native corporations for land purchases, in an effort to preserve them from logging and other Native development projects. Eleanor McMullen, the outspoken tribal leader of Port Graham, extends this to a community self-reflection: "If we cut our trees down or dig up the land what makes us any better than those who created the spill?"[120]

Like the residents of Chernobyl, the people who had a long attachment to land and the water and had harvested their food from them now saw a landscape with a new layer of meaning, the presence of oil, perhaps hidden in marine life that carried visible defects such as strange spots. As Violet Yeston, another member of the Port Graham community, explained, "They say the oil is gone. Where did it disappear? We know it's still there in the ground."[121] After protesting the initial dismissal of these observations and their concerns about damaged wildlife populations, the villages of Tatitlek, Chenega Bay, and Port Graham hired facilitators to coordinate their participation in environmental research that uses traditional knowledge and science. The *Exxon Valdez* Oil Spill (EVOS) Trustee Council, charged with distributing money from Exxon's $900-million settlement with the State of Alaska, has funded such projects as training local residents in sample collection to look for lesions and tumors, seeding clams using local information on best spots, and providing conference opportunities for community members to discuss the best restoration methods. It is now becoming apparent, some years after the spill, that a combination of Western science and traditional and local knowledge is necessary to understanding spill effects and reinvigorating damaged communities that seek self-reliance and healing.[122] These dynamic hybrid knowledges stand against the disaster and its potential smashing of culture.

Public Relations

Throughout the ongoing community struggle for meaning, voice, and compensation, the Exxon Corporation sailed through the turbulence with hardly a dent. In fact, Exxon's tight image control, manipulation of scientific data, and redeployment of easy symbols provided by the national media have served it well in the legal theater. Of criminal penalties of $150 million imposed by a U.S. district court in 1991, Exxon was forgiven all but $25 million because of its "cooperation with governments during the cleanup, timely payment of many private claims, and environmental precautions taken since the spill."[123] Exxon also paid $900 million in civil penalties to the State of Alaska for restoration of publicly owned natural resources. (The state's uses of this money remain controversial, including the construction of the Sea Life educational and research center in Seward, Alaska.) As of this writing, Exxon has yet to pay any of the $5 billion in punitive damages awarded to plaintiffs in 1994, now in appeal. In a separate settlement, $3.5 million was awarded to fishermen, cannery workers, and Natives. Exxon's insurers paid $480 million of its cleanup costs. Thus, although nearly everyone agrees that the human and wildlife communities of Prince William Sound have not been fully reinvigorated,[124] the company has suffered very little in real terms. In 1993, just before a jury awarded the elusive $5 billion to Alaskans who had experienced a massive devastation of their waters and wildlife, Exxon made the most profit of any Fortune 500 company, $5.3 billion.[125] By 1996, its profit had reached $6.98 billion, again more than any U.S. company.[126] If Exxon had suffered a "black eye," as one study said of its public relations performance,[127] why was it doing so well?

One of the big consequences of the *Exxon Valdez* spill was the wholesale shift of crisis management from the practical logistics of containment, cleanup, and compensation to the protection of image. As "corporate problem solver" James Lukaszewski explains, the *Exxon Valdez* created a "robust" public relations crisis management industry because it showed corporate elites that "public perception of how a company handles a problem ultimately matters more than the facts."[128] For Lukaszewski, the *Exxon Valdez* disaster eclipsed the release of methyl isocyanate at Bhopal's Union Carbide pesticide plant, which left thousands dead, a position comprehensible only if damaged image is more important than loss of life. Another disaster consultant, Ron Rogers, argues that "corporations today all operate in a 'crisis environment'" and that the *Exxon Valdez* disaster was a case study in slow response and insensitivity, needing better strategies of image management.[129] Among corporate advisors, the accepted view is not so much that Exxon failed by massively polluting the sound, but that it had erred in its self-presentation because it was "overwhelmed by the emotional magnitude of the event."[130] And Exxon president Lawrence Rawl himself initially agreed with that assessment when he told *Fortune:* "Our public image was a disaster once that ship went on the rocks."[131] Disaster containment is thus applied to company *image,* a signification system that, like a pesticide factory, a nuclear plant, or a spaceship, can be made safe through proper management.

In the virtual milieu of postindustrial capitalism, there are only virtual disasters, and "damage control" is not dirty hands-on cleanup of oil, but careful siphoning of information and corporate metaphor. The new public relations trend is toward "perception management," developed by Burson-Marsteller (B-M), which handled Union Carbide in the aftermath of Bhopal and worked for the Alaska Seafood Marketing Institute to help suppress consumer fears about oily fish after the *Exxon Valdez* spill. As one B-M executive describes it, "We look at strategic ways to manage people's perceptions in terms of hot buttons."[132] The theory behind perception management is that in a crisis, "perception" is out of synch with "reality" and successful managers must bring the two together, asserting control over the internal and external "chaos" created by a disaster. A large-scale disaster creates a crisis of public confidence that must be restored through "shaping and projecting credible renditions of reality,"[133] as Stuart Ewen put it in his history of "spin." Ewen argues that public relations have often been used to protect powerful business interests against popular democratic movements. The large-scale disasters of the 1980s exposed the particular version of reality in which big corporations were engaged in the Reaganomics of technological progress, and fostered a wide cultural critique of dangerous systems. Since then, corporations have accepted the reality of disaster as a problem of perception. As one technical manager for the oil transportation trade explained, oil spills are not disasters but "dramatic events."[134]

Although the 1990 Oil Pollution Act mandated spill drills because of the sloppy and obviously unprepared response to the *Exxon Valdez* accident, disaster drills now often take the form of perception management rather than prevention, cleanup, and evacuation. Company employees and management students enact disasters to test their communication skills rather than their ability to stop environmental damage. For example, one PR firm staged a hypothetical scenario for chemical giant Imperial Chemical Industries in which a tanker crash in Italy sent gallons of chemicals into Lake Garda, killing tourists and local inhabitants. The lesson? How to handle internal communication and the media. The perfectible organization is able to absorb and contain such accidents through the reconstruction of a safe image in the face of the very disasters it perpetually creates and considers normal. Crisis management must be viewed as both external and internal because through these drills, companies maintain control over the perception of employees, gaining their consensus in the lack of real safety as they promote the idea of disaster as ordinary, a communication problem. Thus, Charles Perrow's argument that accidents are normal to complex, tightly coupled systems, and his recommendation that we thus end our reliance on such systems, has been simply absorbed into corporate identity.

Although the corporation's expressions of concern for victims may be met with the skepticism they deserve, the more important corporate goal is to stabilize its identity around the disaster as a containment strategy. Public relations experts maintain that Exxon, which did not hire an external PR firm to handle the crisis, did not handle the media well. According to these critics, who focus only on the first days after the disaster, Exxon's apparent mistakes revolve around Exxon president Lawrence Rawl's failure to appear at the site until 22 days later, Exxon's perceived arrogance in refusing to take blame in its first public statements, and its production of a series of videos that glossed over the spill's effects. Rawl's reluctance to visit Alaska is not surprising considering that in 1984, Union Carbide's CEO, Warren Anderson, was arrested in Bhopal as soon as he stepped off the plane. In his infamous interview in *Fortune,* Rawl said that he realized that "we were going to be up to our butts in alligators right here. I wanted to be able to deal with Congress, as well as operate the best we could around the world."[135] A year later, Rawl, who received a bonus in the spill year, reiterated that his presence wouldn't have made any difference and that he "had other things to do, obviously."[136] He blamed environmentalists for exploiting the situation and said of Exxon's reputation, "we'll win it back, but we're not going to do it by debating on TV with some guy who says, 'You know, you killed a number of birds.'"[137] The refusal to debate with "some guy," suggesting that the company was not interested in public relations even in the act of doing public relations, remained Exxon's position throughout the years of cleanup and litigation.

Rawl's position was consistently echoed by other Exxon managers involved in the spill cleanup. The company's spill response manager, Frank Iarossi, complained, "Exxon is a very, very proud company that has done a lot to protect the environment. . . . The public's reaction is totally irrational."[138] Otto Harrison, who had been brought in from Esso Australia to direct the cleanup, told an Exxon employee magazine, "I don't really worry about the headlines or the politics. I don't think we're suddenly going to be loved by the press or the TV or the radio in this issue. But I do think as time goes on, they'll see evidence of the high technical and operational quality that's going into this [cleanup]."[139] In this way, Exxon presented itself as organizationally immune to disasters such as oil spills and the irrational chaos of public criticism. Exxon's positioning of itself as a rational, disciplined army with scientific and technological expertise ultimately worked. Perception management dictated that Exxon be immune to public perception.

Because Exxon survived nearly unscathed, and by local Alaskan accounts even won the public relations battle, the image makers' argument that the company had "shot itself in the foot"[140] is distorted and self-promoting. In fact, in the view of local residents, Exxon's image campaign was ultimately successful in masking the ineffectiveness of the cleanup that many now agree was futile and even damaging to the beaches' ecosystems. Local residents felt that they were being incorporated into a cynical spectacle over which they had no control. In its chronology of the spill, printed seven months after the disaster, the *Homer News* recounted a series of incidents revealing the local feeling that what Exxon said did not match the residents' immediate experience. On the same day as an Exxon representative declared their beaches safe, Homer residents watched oil wash up into Kachemak Bay. They accused Exxon and Veco of rendering local cleanup efforts useless, taking over volunteers "because we were picturesque."[141] The local view that Exxon and Veco were engaging in a cynical show for the cameras was heightened when reportedly "a video crew from *USA Today*'s TV news program tape[d] workers industriously removing oil from Gore Point and Rocky Point; afterwards, the news crew discover[ed] that the workers had been shuttled to the beaches an hour or two before they were taped, and that Exxon had removed the workers within an hour or two of the time they were videotaped."[142] John Mickelson, from the nearby town of Seldovia, even attempted a citizen's arrest of Exxon spokesman Dan Jones for "conveying false information with reckless disregard for human life."[143] Another resident said, "It's all been just a PR game. They really haven't been concerned about picking up the oil, they are concerned about putting people close to the beach and close to town where they can be seen and everybody will think they are getting the job done."[144] When Veco chairman Bill Allen bought the *Anchorage Times* a year after the spill and ran Exxon publicity photos of "clean" beaches without crediting them, many residents reacted with skepticism.[145]

The cleanup effort, promoted as a "miniature Normandy invasion,"[146] helped shift Exxon's image from wilderness destroyer to a "bumbling, but basically hard-working corporation" that was atoning through expenditures of capital.[147] Throughout the spill cleanup, one of Exxon's strategies was to create an image of itself as an army besieged by hostile forces, shoring up internal loyalties. In Valdez, center of cleanup operations, Exxon appropriated buildings and grounds and placed armed guards around its headquarters there. Local residents perceived Exxon's presence as a military occupation.[148] Exxon created and encouraged this view of its cleanup workers as an invading army. In a PR video for shareholders, Exxon described its "landing force in a hostile environment," engaged in an "all out amphibious assault."[149] And in its internal employee publication, Exxon's workers were portrayed as technically adept soldiers, lonely for their families, risking their lives, surviving the harsh conditions of working in snow and rain in a remote and distant land.[150] Like soldiers, they developed a "bunker camaraderie" and called their brief Alaska assignment, a "hitch" or a "tour."[151] The military ethos extended to corporate communications: Oil company employees and subcontractors were required to sign agreements saying they wouldn't speak to the media.[152] Complaining that the company had canceled briefings and press conferences after a National Transportation Safety Board investigation in April 1989, one reporter wrote of Exxon's Valdez headquarters, "Suddenly you couldn't get near the place. There were armed guards all over the place. It was like a fortress."[153]

Exxon's enemy, around which the company coalesced in the postdisaster milieu, was never clarified, and at various times took the form of oil, nature, the State of Alaska, the public, and the media. Charles Wohlforth, who covered the spill for the *Anchorage Daily News,* later described the popularity of military imagery among journalists: "Writers used the war metaphor as if it were an express train to the truth and no other was leaving the station. In their articles, workers on the beaches became soldiers landing at Normandy or retreating from Dunkirk. I was guilty of comparing Valdez to Saigon. But the war metaphor didn't really fit. Who was the enemy? Not the oil. It didn't fight back, it just sat there. It wasn't trying to prove a point. We were."[154] Even used in a critique of Exxon's practices, as in the comparison of Valdez to Saigon, the war metaphor could only serve Exxon's self-conception as an army in a hostile land, a nation unto itself. Because it had positioned itself as a rational, disciplined technological agent in the midst of irrational forces, such external critiques mattered little. In the immediate aftermath of the spill, internal cohesion was most important considering the growing demoralization among Exxon employees.[155] While local people accused Exxon workers of having an asphalt jungle consciousness and lacking a sense of place, this simply served the company's own image reconstruction. A year after the spill, *U.S. News & World Report* called the spill "the disaster that wasn't," and declared, "contrary to a public perception fueled by one-dimensional television images and a furious environmental community, Prince William Sound is no longer an ecological disaster zone."[156] In 1994, after plaintiffs won their $5-billion settlement against the company, *Forbes* portrayed Exxon as a hardworking company beset by litigious interests trying to gouge "a chunk of Exxon flesh," Rawl as a "babe in the woods," and Exxon's shareholders as rape victims.[157]

These adaptations and reversals of national media images of the spill made Exxon out to be the victim, but Exxon itself was more interested in constructing an identity that laid claim to scientific objectivity, a transcendent position that would allow it to rise above the disaster. Exxon's science was a slow and deliberate campaign not only to deny that the oil had any negative effects but to argue that it had actually stimulated the ecology of the sound. Drawing its truth claims from scientific experts, Exxon created a vision of fertility, biodiversity, and evolutionary change within a wide global context that dismissed specific damages to local wildlife populations. Exxon claimed that its scientists took a more appropriate "global" view of events[158] and were more objective than government scientists, who had been affected by "potent images of dying animals."[159] Therefore, Exxon converted the critique that the company had reacted in an insensitive and arrogant way into an argument that it was simply being objective and rational.

However, that claim does not mean that Exxon was, or is, a rational entity, even though it may have presented itself as operating through a strictly scientific and technological frame of reference. In his study of the events that led to the Bhopal accident, Paul Shrivastava argues that "corporate frames of reference are typically rigid and scientific, built from objective technical data and rational cognitive maps."[160] This, he says, makes the corporation seem callous and unconcerned about the human and environmental costs of disaster. However, although the public image projected by the corporation may be of a single rational entity, it is simply that—an image. In the postdisaster milieu the aim is not to produce good science but to reconstruct a fractured image, because a disaster exposes the gaps and slippage in the rationality of the organization. Exxon's discourse on postspill ecology was denatured and self-referring, creating an imaginary of the corporation itself through the metaphor of Prince William Sound's miraculous resilience.

Exxon's communication experts deployed science to restore an image of the corporation as a life-giving cornucopia, a stimulator of natural processes. Exxon's public relations brochures, advertisements, and press releases used a highly vague and emotive language disguised as scientific conclusion. According to Exxon's environmental biologist Andy Teal, Exxon's key media phrases "were developed in the field and then adopted by headquarters."[161] In April 1990, a year after the spill, Exxon brought in three scientists from the United Kingdom to survey Prince William Sound and the Gulf of Alaska: consultant Jenifer Baker, emeritus zoology professor Robert Clark, and team leader marine biologist Paul Kingston. These experts spent two weeks around the site, visiting fifteen locations, vaguely concluding from this cursory examination that the beaches had a "generally favourable appearance."[162] They returned in 1991 for another field observation in which they concluded that the sound was nearly recovered. Their findings were transformed into attractive pamphlets, full of glossy coffee-table photographs with captions such as "Fresh grass shoots sprouting from sediment containing residual oil"[163] and "Mature rockweed at Sleepy Bay, La Touche Island. The ends of the fronds are swollen with spore-producing structures."[164] The emphasis on an "abundance" of mussels and barnacles, an "explosion of limpets," "thriving" kittiwakes, "abundant" and "thriving" worms and shellfish, rapidly reproducing otters, sea lions basking on previously oiled rocks, and record harvests of salmon all suggested not only that the oiled areas were recovering, but that the oil spill had actually been good for the environment. The language of fertility and bounty was not scientific, not based on any apparent scientific evidence other than a mostly casual, selective observation filtered through the lens of image recovery. Statements by other "experts" who ran Exxon's bird and otter recovery centers were incorporated into pamphlets that boasted of a "healthy, thriving eagle population" with access to "ample food sources" in a "saturated" environment, and "plentiful otters" with "ample stocks of shellfish."[165] The repetition of a few stock adjectives appeared both in these technical report summaries and in Exxon advertisements, which boasted of a "thriving" wildlife in the sound.[166] Thus the language of science became the language of public relations.

As in environmental studies of the new Chernobyl wilderness, the landscape was restored in a discourse of plenitude. But oil is not radiation, and does not carry the same menace of a hidden, hostile force lurking under the pastoral reclamation. Instead of an adventurer in a postapocalyptic world, Exxon presented itself, and its technologies, as an intimate partner of nature. In one Exxon postspill brochure, featuring artists and tourists attesting to the beauty of the sound, Exxon engineer Nick Martinez professed how "good" it felt to be part of the cleanup: "Nature takes care of itself real well, but we speeded up the process of getting the shorelines back to their natural state. . . . It's obvious that we and nature are doing something right."[167] Although winter wave action did the most to clean up beaches in the spill areas without Exxon's help, Exxon's public relations relied heavily on the promotion of bioremediation, the application of phosphorus and nitrogen to increase bacteria that convert hydrocarbons into carbon dioxide and water. Or, as Exxon put it to its shareholders, bioremediation is "fertilization," the "encouraging [of] indigenous bacteria to consume oil," thus "enhancing natural processes."[168] In this corporate vision of environmental relations, oil consumption was naturalized, native species were well fed, nature was fertilized, and continuing profit was ensured. In 1997, Exxon's chief environmental scientist on the spill, Alan Maki, was still pounding the message home in public lectures: "Mother Nature's natural cleaning" had been "complemented" by Exxon's "fertilizer."[169] Bioremediation was a public relations coup because many promoters later pointed to the spill cleanup as the first large-scale use of the technique.

The discourse on bioremediation subsumed Exxon's problems with Alaska Natives, fishermen, oil consumers, and ecofeminists and other environmentalists into a simple set of optimistic messages about benign corporate technologies,[170] offsetting criticism that Exxon's hot water washing techniques had sterilized the beaches. Although Exxon publicly claimed that its bioremediation agent, Inipol EAP 22, was like the fertilizers "you might use on your lawn,"[171] its warnings to workers spraying the beaches told a different story. In an Exxon Alaska Operations safety bulletin, Inipol was compared to pool cleaners, stain removers, wax removers, and carburetor cleaners, and workers were warned that it might cause severe eye and skin irritation and, with prolonged contact, damage to red blood cells. Inipol was hardly as benign and natural an agent as Exxon wanted its investors and consumers to believe.

Exxon's public relations campaign transformed scientific discourse into key terms and recycled media images to create a new global corporate identity, absorbing the disaster through redeployment, reversal, and denial. Promising an uninterrupted bounty of consumer goods made from petroleum in harmony with a fecund environment stimulated by oil and chemicals, Exxon's discourse was neither ample nor plentiful, keeping scientific data secret in the interests of litigation, stripping down language to "hot buttons," and ignoring local complexities in globalizing frames of reference. This reductionism was designed to protect Exxon's interest in the Arctic National Wildlife Refuge and its shifts in investment to overseas ventures that supply most of its earnings, including its deal with the genocidal Indonesian government to drill in the Natuna oil field. A discourse that emphasized the stimulation of native species to become biological factories for nutrient oil consumption was certainly useful to an oil company with vast overseas investments. Exxon's Save the Tiger campaign, often criticized as "greenwashing," was nothing in comparison to this subtle "engineering of consent"[172] that carried the weight of scientific truth. Meanwhile, the common murre, the harbor seal, the harlequin duck, the marbled murrelet, the pacific herring, the pigeon guillemot, the pink salmon, the sea otter, the sockeye salmon, and the damaged subtidal and intertidal organisms[173] have given no consent to their injuries or the devastation of their local populations. Such a consent to the terms of disaster can never be engineered.

5. The Bhopal Effect

No More Hiroshima

In the year following the Bhopal disaster, a release of chemicals that left more than 6,500 people dead and another 50,000 people partially or permanently disabled, survivors held a series of public protests for compensation and retribution. Six months after the disaster, they blocked a state government office, demanding relief monies, and were beaten by police with truncheons. At the first anniversary, they protested outside the gates of Union Carbide (India) Ltd. (UCIL), despite prior arrests of activists planning to demonstrate. An effigy of Union Carbide CEO Warren Anderson, with thirteen heads, was burned and beaten, an incident widely reported in the U.S. press for its conventional vision of Third World violence, anger, and hostage-taking. Rarely given any public forum for their own representation of the disaster, the protesters used the available means: the prayers for the many dead, the making of personal monuments, the collective call to remembrance, and the angry public exposure of their own weakened bodies marked by disaster's scars.

As director of the Indian Law Institute Upendra Baxi says of the victims, the chemical disaster "produced toxic impacts on their docile bodies" while the courtroom battle with Union Carbide and the Indian government for compensation "accentuated their agony and erased them from history."[1] Protesting their abstraction into undifferentiated legal "categories and aggregates," Baxi calls for a "concrete" and "evocative language of human suffering" that moves beyond mere rhetorical device.[2] As Elaine Scarry suggests, pain, which has no object, is a "shattering of language" into inarticulate cries, an interior, closed, subjective state. And the act of drawing pain into language can be a project of both torturers and healers, and thus has great ethical consequences.[3] Scarry's focus on the domestic technologies of torturers—for example, the conversion of soda bottles into bludgeons—resonates in disaster narratives, which often feature this evil transformation of the world of familiar but unstable objects. Unable to be reduced to an objective accounting or public relations, suffering takes on the hard currency of a reality that is caused by, but ultimately resists, technological systems. Whereas postindustrial cultures offer an elusive happiness through pleasurable interfaces with the exported global technologies of the simulacrum, the suffering of disaster victims serves as a reminder of its limits. The rigorous suppression of voices from Bhopal that acutely bear the traces of that suffering is present not only in the courts, but in other governing bodies, in the international media, and even in the streets, where it has taken a violent form.

The erasure of Bhopal's victims began in the necessary failure to keep official records of the dead, in a world where even genocide is subjected to a minute accounting process, leaving an eerie paper trail. No one knows how many people died at Bhopal or how many were injured because of the chaos of the event, during which many people fled and died in other places and the dead were given mass cremations and burials or put in the river.[4] Some Indian observers suggest that local government officials buried humans and animals together in mass graves to suppress the number of deaths in the interests of social order.[5] As the UCIL shift timekeeper later remembered, when the plant superintendent asked him how many workers had fled the accident scene, he had to explain that they did not stop to ask for pass cards.[6] The resistance of mass disaster victims to an exact accounting makes it difficult for institutions to completely absorb them into technorational narratives.

This has had unfortunate consequences in the courts. As sociologist Veena Das points out, the survivors were constructed there as "malingerers, irresponsible litigants and simple frauds."[7] Because no firm number of casualties emerged and because chemical injury is not completely understood by science, their suffering could easily be denied legal validity. Furthermore, Das

The Bhopal Gas-Affected Women Workers' Organization marching in protest.
(Photo by Amrita Basu. Courtesy of Amrita Basu)

suggests that the anguish of Bhopal's victims was subsumed by these legal demands for scientific evidence so that it became merely an "ornamental" display in the courts: "The more suffering was talked about, the more it was used to extinguish the sufferer."[8] Ultimately, only $470 million was awarded to the victims in an out-of-court settlement with Union Carbide, an outcome that is almost universally regarded outside the U.S. business press as at least inadequate, if not unjust.[9] Distribution of this settlement to victims, many of them illiterate, has been delayed and has encountered disputes over claims, mostly revolving around birth, domicile, and medical records.[10] The claims process is intricate and given to official corruption, requiring survivors to offer bribes, spend money on photocopying, carry elaborate documentation, and recount painful memories of the traumatic event that ultimately have no bearing on the outcome. Like Foucault's plague town, Bhopal has become the site of surveillance, observation, recording, and writing to contain the disaster, always inadequate to the tasks of healing and redress.

In the United States, where the voices of victims might have swayed public interest against Union Carbide and forced a larger settlement, they were almost entirely silenced in the media analysis, which focused primarily on technical causes of the disaster and on Union Carbide's economic survival. Bhopal's victims were present in the U.S. spectacle primarily through photographs that accompanied the news. It was the camera that brought the spectator into Bhopal, to witness the dead lying in the streets and the survivors carrying their dead children covered by cloths or huddling together, their eyes protected by white gauze circles. Although such photographs provoked sympathy and a sense of "shared vulnerability,"[11] their circulation commodified the event and sliced it from the subjectivities of the victims, making even the survivors appear passive and frozen in an eternal moment, in the moment of death.

Because of their otherness, like the birds and otters of Prince William Sound, Bhopal's silent victims became sites for analysis of technological death—and technological manipulation of death—in a way that the familiar, motherly body of Christa McAuliffe could not. Pablo Bartholomew's photograph of a Bhopal child victim, which won a 1995 World Press Award and now appears on an environmentalist Web page with a brief factual account of the Bhopal accident, is a carefully composed totemic image, a death mask, rising from the earth as if from an archeological dig. Viewers are asked to perform an interactive archeology of the image, selecting it from other photographs, deciding how the child died, observing the burial, analyzing who owns the hand that nearly touches the face with such intimacy. They can click on the image to enlarge it for a closer examination, download it, and reproduce it, fingering it themselves. This level of abstraction from the living context, and the objective analysis and technological manipulation it allows, give an illusion of intimacy while placing any suffering at a safe and controllable distance. Furthermore, a photograph of the same burial, with a nearly identical arrangement of elements, appears as an illustration to an article in *India Today,* suggesting that at least two photographers were gathered at the sight composing the same image.[12] That photographs of Bhopal's dead would eventually appear in the *Faces of Death* shockumentary video series suggests the ease with which such images become clearly pornographic, used solely for shock, interchangeable with other images of decapitations, suicides, car bombings, and autopsies. The

Award-winning
photograph of a
Bhopal child's buri-
al. (Photo by Liai-
son/Pablo Barthol-
omew. Courtesy of
Pablo Bartholomew)

infinite reproduction and malleability of photographic images absorbs
technological violence into the continuous media production of violence
without perpetrators.

Although the disaster and its suffering necessarily stand beyond the limits of discourse and photographic representation, acts of survival do not. The
healer's task is not to accurately expose suffering, but to acknowledge the
depth of its silence and to hear and make and support narratives of sustainable survival for those who have been harmed. There is no returning to an
original moment of disaster, but the stories told about that moment provide
a necessary means for constructing lives and communities under new conditions. The public censorship of survivors, so common to media disaster
spectacles, is ostensibly justified on the basis of respect for their suffering, but

suppresses that telling. And the absorption and denial of their stories cut off the authority of their experience.

The voices of survivors, difficult to find in any case, describe that experience in a steady, factual manner, with attention to time and everyday circumstance, as if the dread and terror cannot be spoken fully. Survivors remember getting out of bed and having a glass of water before realizing that something was wrong with the air. They recount a strangely building pressure in the lungs until suddenly they were choking. Nearly every account mentions that the gas was like the smell of burning chilies, a mundane detail that has become an integral part of a shared moment, repeated again and again in the solidarity of the survivors' experience. The narrative does not seek to expose the horror to a hearing, but to establish the authority of witnesses to speak on their own and others' behalf and to bind them to a continuing life history, shared by a community of sufferers:

> The day before the gas leaked was a Sunday. My friends and I were playing in the evening and after that we watched a movie on the television. I must have gone to sleep around 9 P.M. It was cold and I had covered myself with a rug. Sometime in the middle of the night I heard a lot of noise coming from outside. People were shouting, "Get up. Run, run. Gas has leaked." My elder brother Jawahar got up and said "Everyone is running away, we too must run." I opened my eyes and saw that the room was full of white smoke.[13]

Another account of the event, by former UCIL plant operator T. R. Chouhan begins,

> On the night of 2nd December 1984, I was home asleep with my wife and two small children. I first knew there was a problem when fumes of white gas filled the house and we began to choke. The gas itself announced its presence, not any alarm system or public announcement. When our choking continued to increase and our eyes began to burn, I decided that we must leave. Outside there was total hysteria. As I did not own a vehicle, we traveled by foot; my wife was carrying our son and I was carrying our daughter. . . . The streets were full of fleeing people; unable to breathe and vomiting, they continued to run.[14]

Having lost his parents and five brothers and sisters, Sunil Kumar recounts, "The neighbour's shouting woke us up, and all of us ran down the alley. There were thousands running. I was coughing, and my eyes were stinging. People were falling in the road. I didn't look back because I thought my family were still running behind me. But they had fallen, too."[15] And Abdul Jabbar Khan, who went on to become the head of a Bhopal survivors' activist organization, remembers his return to Bhopal after fleeing from the gas: "At Habib Ganj, which is some 12 kilometers from town, I began to see dead bodies on the roadside. I will never forget that sight; it had a powerful effect on me. My sister lived in Kazi Camp. I went there and helped her and her two children leave."[16] Others describe the experience of helping others despite their numbed state: "I covered my face with a muffler. I did not feel the irritation of the gas. As I moved towards my house, there were thousands of people lying on the road, like dead fish on a beach. I started helping people by massaging them or helping them get up."[17]

Survivor Brojendra Nath Banerjee, who did printing work for the UCIL plant and managed to seal himself and his family in his house to escape the gas, recounts stories of people tearing off their clothes from the intense heat of their metabolic changes, leaving their sandals lying in the streets. Some returned to Bhopal to find their spouses and children packed in mortuaries and tagged with numbers, sometimes locating them days later through pic-

Abdul Jabbar Khan speaking to gas survivors.
(Photo by Amrita Basu. Courtesy of Amrita Basu)

tures of the dead posted on the walls of the city. Banerjee writes that they sometimes forgot to weep: "The nightmare sinks into one's consciousness, one struggles to keep the images out, to anaesthetise oneself."[18] Afterward, many survivors reported that they had an intense fear of enclosed spaces, of the wood in their houses that might have absorbed chemicals, of fluttering curtains, of the very air they breathed.[19]

This tone of quiet witnessing, from the point of view of the survivors, stands in contrast to the dramatic narratives of urban holocaust appearing in the U.S. news media in the first days after the disaster. Here, where every violent event has its Ground Zero, nuclear war provided a dramatic, conventional image, disconnected from a culpable agent, from Union Carbide, suggesting that the gas was a mysterious and ineffable force from nowhere, the numinous "strange-god"[20] of technological nightmares. As the *New York Times* reports, "It began without warning in the dead of night, while the vast and crowded slums of Bhopal, India, lay in slumber, dreaming the troubled dreams of want and hope, heedless of the danger in the wind scything over the silent metropolis."[21] *Time* describes the aftermath as an "eerie science fiction nightmare," an urban post-apocalyptic environment in which soldiers and "volunteer vigilantes" picked up corpses, and "vultures and wild-eyed pariah dogs roamed through the piles of rotting flesh, feasting."[22] References to nuclear war are rife in these accounts, the rumors of atomic attack: "Saha made his way to the railway office, only to find the stationmaster slumped over his desk. For a moment, he thought that an atom bomb had hit Bhopal."[23] In 1984, deep in the Reagan era of "Star Wars," first-strike capabilities, and biblical prophecies applied to foreign policy, the discourse on the Bomb and the end of world through plague had a powerful resonance.

A few years later, journalist Dan Kurzman elaborated this theme in a book-length dramatization. It begins with an evocation of Hiroshima, and the spectacle of an "amorphous gray cloud . . . waiting to vent its rage on a world either scornful or ignorant of the powers it is now free to reveal."[24] This personification of the gas cloud, accompanied by the apocalyptic language of hidden power and revelation, carries the conventions of nuclear discourse. The narrative unfolds through several characters, such as Union Carbide CEO Warren Anderson, a water woman who worked at the railway station, a marijuana-smoking holy man, a plant worker, and a beautiful young widow. Like the old lifeboat narratives in which several representative types provide lessons in moral character and the tension comes from guessing who will live and who will perish, the book follows these residents through the disaster and its aftermath. Rather than allowing the actors to tell their own stories, Kurzman robs them of their own mental depths and provides them with inflated, impossible thoughts proving certain moral points about sin and redemption. For example, Warren Anderson is supposed to be haunted by the ghosts[25] of Union Carbide's past environmental record, while the night-shift worker is supposed to suddenly realize that his company is evil: "Ali was becoming delirious as he scrambled forward searching, calling, hoping. The dreadful noise of the siren that he had hardly ever noted before pounded in his head like the laughter of the devil."[26] Because everyone in the story is treated with an even hand, as having similar emotional states all provided by the author, the narrative is a gross appropriation and erasure of suffering under an amorphous cloud that seems to descend under the apocalyptic will of a nuclear-age heaven.

Such depictions of Bhopal presented a worst-case scenario that, like a postapocalyptic film, had not yet been fulfilled for its viewers. But for the thousands of dead, the disaster *was* the end of the world, not merely a rehearsal. Survivors had a closer claim to Hiroshima because of the mass death caused by a technological agent. The mayor of Bhopal explained, "This will never be written off from our minds—just like the people of Hiroshima and Nagasaki still remember."[27] And for survivors in their long struggle for healing and compensation, comparisons between Hiroshima and Bhopal were politically charged, signifying the deadly incursions of a U.S. military industrial complex supported by the Indian government. At the first anniversary, as was widely reported in the news, protesters painted "No More Bhopal, No More Hiroshima. Save the World for Future Generations" on the UCIL factory walls, accompanied by a skull with dollar signs in its eyes.

Comparisons of the city to a gas chamber, later bolstered by discoveries that the gas cloud had caused cyanide poisoning, also evoked a programmatic technological genocide. This comparison was further reinforced by charges that the UCIL plant was engaged in chemical weapons research, first sounded by Romesh Chandra from the World Peace Council[28] at a conference in Bhopal, with later observers suggesting that the disaster had been a deliberate experiment.[29] Much of the evidence involved the activities of a Union Carbide toxicologist, rumored to be working for the U.S. "Chemical Defence Department."[30] According to Ashis Gupta, he had once worked at a British biological and chemical warfare laboratory and was rumored to have carried out blood tests and lung autopsies on survivors without their consent.[31] Union Carbide's initial statement that methyl isocyanate (MIC) was no more dangerous than tear gas, and reluctance among government and medical personnel to recommend treatment with sodium thiosulfate, a cyanide antidote, further suggested a cover-up. Organs from autopsies are rumored to have mysteriously disappeared, medical records secreted off. The hospital under construction through donations from Union Carbide is said to be a front for further biomedical research.[32] Theories linger among survivors that the plant was a cover for chemical and biological weapon manufacture, that the disaster had attracted chemical warfare experts who used the victims as guinea pigs.

Whatever truth there might be in these allegations, the suggestion of conspiracy provides an alternative version of events that highlights the intent and culpability of Union Carbide, its international networks of power and influence, its collusion with the state, the potentially genocidal aspects of complex technological systems, the callous disregard for civilians, especially the poor, and the manipulation of science. This version stands against the U.S. media's wholesale protection of Union Carbide, including support of the company's theory that a disgruntled employee had sabotaged the plant and brought the disaster onto his own community. Although the media has suppressed any investigation of chemical weapon research, it has paid a great deal of respectful attention to Union Carbide's saboteur theory, which transformed the disaster into workplace politics. But for survivors, the implications of a war waged against them have provided a broad, international purpose, in solidarity with Hiroshima's *hibakusha,* to prevent further incidents of massive technological violence by calling attention to their own scars.

As part of the silencing of survivors, their activism has been portrayed as the result of outside infiltration by self-serving environmentalists or communists preying on their fears.[33] It has also been suggested that global concerns over the chemical industry are the purview of these "issue-entrepreneurs," whereas the victims themselves are more interested in immediate compensation.[34] As is typical in disaster discourse, survivors are seen as unable to speak for themselves, locked in debilitated bodies, frozen in the moment of pain. However, considering the outspoken books written by survivors and the frequent and long-lasting protests more than ten years later, it is apparent that many are deeply invested in environmental, social, and economic change. Nine years after Bhopal, the convenor of the activist organization Bhopal Gas Peedit Mahila Udhyog Sangathan (the Bhopal Gas-Affected Women Workers' Organization, or BGPMUS), Jabbar Khan, appealed for aid for Bhopal's victims and wrote of the Indian government's acceptance of international trade rules: "It takes little to imagine how this would lead to the setting up of hazardous factories, exploitation of the country's natural wealth, ruination of the health and life of the people, and ecological devastation. Friends, in these times we have to join forces to fight the unholy collusion between the government and foreign capital that is causing destruction in the name of development. The Bhopal gas disaster . . . is a burning example of this antipeople collusion."[35] Shahazadi Bahar, another member of the BGPMUS whose eyes and lungs have been permanently damaged, explains that she initially joined the organization in the selfish hopes of getting a job, but then began to fight for a cleaner environment as well as better medical treatment, compensation, and punishment of Union Carbide.[36]

Political scientist Amrita Basu, who studies Indian women's grassroots activism, has chronicled the activities and motivations of the BGPMUS, mostly made up of poor Muslim women who organized to protest the shutting down of sewing centers used for rehabilitation of survivors and later broke into the Union Carbide office in Delhi after the legal settlement, breaking windows and painting slogans on the walls. She argues that their continuing political battles have been waged not only against Union Carbide and the national government, but also against the local right-wing Hindu government's discriminatory practices, such as the sudden eviction of slum dwellers to the outskirts of the city, an attempt to eliminate gas survivors from sight. Furthermore, these women are particularly angry about the gas damage to their reproductive organs in the face of a political movement that wishes to annihilate them.[37] Mother Theresa further politicized their bodies when she used her visit to Bhopal to speak out against abortions worldwide.[38]

Women's and children's bodies feature heavily in the discourse on Bhopal, medical studies, international media accounts, and protest literature. Over 50,000 Bhopal survivors are disabled, suffering from weakened immune systems, severely scarred lungs, blindness, and loss of speech and memory. Whereas only one study of men's reproductive health has been conducted, involving the sperm of only 18 male survivors, women's reproduction has been the subject of several studies, which found an increase in spontaneous abortions, stillbirths, diminished fetal movements, menstrual disturbances, pelvic inflammatory disease, endocervicitis, menorrhagia, and suppression of lactation.[39] C. Sathyamala argues that the first survey of gas victims by the Indian Council of Medical Research treated women only as reproducers and ignored their other ailments.[40] In the transitions to technoculture, with its fearful possibilities of supplanting biology, women are seen to carry the continuity of humans against the sterile and sterilizing machine, so their bodies are of particular interest in exposing, healing, or ignoring the ravages of technological disaster.

Pregnant women and their stillborn fetuses and genetically damaged offspring provided both an emotional locus and an analytical site for examining the gas's effects. In protest literature, fetuses are described as having been scorched in the womb, babies born with no hands, holes for eyes, blue and blistered and burned skin.[41] A collection of firsthand accounts of UCIL plant workers begins with a brief statement from a member of the BGPMUS who describes her three spontaneous abortions: "They were all born dead. All with black skin like the color of coal and all shrunken in size. The doctors never told me why such things are happening to me."[42] Although the U.S. media routinely reported in statistical language that stillbirths had occurred, it downplayed any congenital damage. Sheila Tefft, reporting for the *Chicago Tribune* at the first anniversary of the disaster, described a gas survivor in a clinic holding her new son in her arms and claimed that the "one bright spot is a low incidence of deformity among children born in the last year."[43] The media's language of statistics and "low incidences" exercised a control over violent technological disruptions of biological continuity, so graphically represented in the survivors' accounts.

Women's own deployment of their bodies in public spaces, in marches and protests, at the very real risk of being arrested, beaten, and raped,[44] is a visible rejection of that reduction. They have also educated themselves in their own condition. In 1985, after a campaign on women's health sponsored by activist organizations, women gas survivors refused to accept statements by government health officials that slum residents could not be educated to bring their urine for cyanide testing. Any government acknowledgment of cyanide poisoning was politically charged. At a public health meeting, 150 slum women turned up, waving bottles of urine.[45] In its collective expression, the evidence of the disastrous body, still leaching chemicals, held an undeniable authority in the face of official rhetoric.

Kiss a Carbider

A few years after the Bhopal disaster, two management professors interviewed executives at a multinational Fortune 500 company they called "Chemco" in order to assess the impact of the Bhopal disaster on its operations. Worried about the "Bhopal effect," "Chemco" managers saw the disaster as a "signal to the public" that their operations might have the potential for similar "Bhopal-type vapour cloud events."[46] They sent out a questionnaire to all their plants worldwide, asking for assessments of their hazardous chemical inventories and evacuation plans for surrounding communities based on worst-case scenarios. According to "Chemco" managers, safety personnel then recommended improved evacuation plans and elimination of unnecessary onsite storage of hazardous chemicals. One plant subsequently held an evacuation drill that attracted only 500 participants out of 15,000 nearby residents, despite the efforts of Boy Scouts who went door to door handing out information.

This managerial story of post-Bhopal "Chemco" was typical of the chemical industry's response to what it perceived largely as a public relations problem, not only on the international scene, but also in local communities. According to the management researchers, "Chemco" had previously focused its safety concerns on chronic everyday leaks of carcinogens, but shifted its attention to sudden catastrophic events because it feared public scrutiny, especially in terms of new government regulations. News from Bhopal transformed the discursive formation of the chemical company, both in the production of internal safety documents and questionnaires to employees and in the unprecedented output of public service announcements and solicitations of community involvement in staged disasters.

The deflection of attention from the horrific deaths at Bhopal to the chemical industry's image was supported by the U.S. media and business analysts.[47] Personal injury lawyers, who went to Bhopal within a few days of the disaster, became popular scapegoats, representing callous greed, often described as vultures swooping down on the injured populace. On the other hand, most discussions of Union Carbide's role were concerned primarily with correcting its performance and saving it from dissolution, rather than with the economic, social, and medical consequences for Bhopal's victims. Unlike Exxon's supposed public relations debacle during the oil spill that would occur five years later, Union Carbide was presented sympathetically in the U.S. national and industry press and was ultimately perceived to have handled the Bhopal crisis well, with the requisite contrition and self-preservation. In essence, however, the oil and chemical industries deployed similar strategies of image reconstruction, often called greenwashing, as management experts declared an age of "Mega-crises" that threatened the "existential core" of corporations and the "legitimacy of an entire industry."[48]

The media's attention to the fate of Union Carbide shifted focus from the tragedy in Bhopal to the preservation of U.S. capitalist interests, but it also revealed fundamental contradictions between the technorational image of the corporation and the emotional urgency of the disaster spectacle. Major news stories about the deadly effects of MIC and residents attempting to flee the gas cloud that burned their eyes and scorched their lungs were accompanied by sympathetic descriptions of Union Carbide president Warren Anderson and the response of employees at the corporate headquarters in Danbury, Connecticut. In *Business Week,* which described Union Carbide as "fighting for its life," Anderson initially described his reaction to news from Bhopal: "I almost felt that if I'd go back to sleep and wake up all this would disappear. It's a shattering experience."[49] *Fortune* noted that he was "home in bed with a bad cold"[50] when he heard the news.

Under the advice of company lawyers and PR consultants from Burson-Marsteller, Anderson flew to Bhopal within a few days of the disaster and was detained in the Union Carbide guest house for six hours by the Indian government. In a typical frontier drama of captivity, the Indians held hostage the white man who had "struck off alone" rather than "hiding in the corporate bunker."[51] The national press featured the "personal trauma" of the former "football letterman"[52] whose trip was called "courageous and humanitarian" by Jesse Werner, former chairman of GAF, another Fortune 500 chemical company.[53] Burson-Marsteller senior executive Jim Lindheim explained that Anderson's brief arrest was a public relations coup because it called attention to the corporation's demonstrations of concern in the wake of an irrational response in a distant land: "People have this image of flaky Indians. It isn't really true, of course, but, you know, teeming masses."[54] Anderson served as a symbolic bridge between U.S. headquarters and Bhopal, a connectivity that was both a confirmation and plausible denial of Union Carbide's legal responsibility. After this visit, during which Indian officials rejected Anderson's offers of crisis management, including a visit to the site and a meeting with Rajiv Gandhi,[55] Union Carbide began a legal battle to divest itself of any connection to its Bhopal plant. Thus Anderson's visit remained in the realm of a brief and fleeting connection, without tangible obligation, without solidity or durability beyond the past satisfactions of profit and the momentary exigencies of crisis. The thrust of Union Carbide's crisis management was to launch a lifeboat according to the logic of the commons, denying its networks of subsidiaries and divisions, avoiding implications of shared goals, technologies, and management that might lead to legal responsibility.[56] This resulted in a continuing war of ownership over Anderson's body, including efforts by the Indian courts to extradite him for culpable homicide and corporate attempts to shield him, concealing his address and pleading his advanced age.

Anderson returned to the United States, reported London's *Financial Times,* "with the poignant observation that the rest of his working life would be devoted to sorting out the Bhopal problem."[57] *Industry Week* worried broadly about "Bhopal's shattering effect on the executive psyche."[58] And when the *Challenger* exploded, Morton Thiokol CEO Charles Locke and Anderson called each other at the same time, "almost telepathically" as *Business Week* reported, and Anderson advised Locke to avoid exhaustion by sleeping on a cot kept in his office.[59] The *New York Times* asked "eminent" psychologists to comment on the normally "shy" Anderson's response to the disaster. They found him to have "an unusually healthy and constructive manner of dealing with his own reactions," helped by his distance from the "worst of the carnage."[60] Along with the media attention to psychological effects on the executive mind, Anderson's body, as the representative corpus of the corporation, was publicly examined for its symbolic implications, as it was restructured to manage the crisis. Commenting on Union Carbide's divestitures in the wake of Bhopal, *Chemical Week* observed that Anderson "had lost weight and his trousers were loose. Belt tightening, as it turned out, was soon to become the rule for the entire corporation."[61] As in other disaster narratives, Anderson's story played out the moment of rupture, the shattering of an ordinary night, the revelation of unforeseen dangers, the sudden visibility and fame, and a spiritual identification with a new peer group, executives in crisis.

Balancing the experience of all "stakeholders" and using these familiar narrative conventions, the industry asserted Anderson's crisis as equal to the suffering of Bhopal victims, described in *Fortune* as "crowds of men, women, and children scurrying madly in the dark, twitching and writhing like the insects for whom the poison was intended."[62] In contrast to this portrait of an unruly, crowded postapocalyptic land, Union Carbide's headquarters in Danbury was described as a quiet "architectural showpiece secluded on a wooded ridge,"[63] suddenly silenced by the news. MIC production was halted at Bhopal's sister plant in Institute, West Virginia, where at least one worker took off his Union Carbide jacket in shame. Union Carbide employees were said to be weeping in the cafeteria at Danbury.[64] The "happy shop" where "jokes were freely exchanged in the corridors" had "changed dramatically."[65]

But any emotion felt for the people of Bhopal became a performance of crisis management, carefully circumscribed within the parameters of industry self-interest. As Union Carbide's director of corporate communications, Robert Berzok, told a conference on environmental management, the company attempted to make an "emotional connection to the public" through videotapes featuring Bhopal's dead children.[66] Anderson's press conference upon his return from India, during which he discussed his "shattering" experience while promoting his company's safety record, was broadcast to employees by closed-circuit television. In a letter on the cover of the company magazine, Anderson declared that "we're showing the world that Union Carbide is, indeed, a family, one with a heart and soul in all the countries where we do business."[67] The emotional life of Union Carbide's workers, which might have generated some active feeling for the victims of Bhopal, became a commodity in the symbolic exchange, a commodity released in order to demonstrate its manageability within the Union Carbide "family."[68] Thus, in the same letter Anderson asserted that there was no "lasting damage" to the people of Bhopal, and ended by pro-

moting his company's "plans" and "strategies" that would help it achieve "higher levels of performance." Several years later, the corporate director of safety
at the time of the accident recounted that while he still wept at the thought of
the Bhopal disaster, he wanted "future generations of Carbiders" to "respect
it" rather than be "haunted by it."[69] Whereas the gas victims of Bhopal suffered
from damage to their own reproductive organs, the reproductive continuity—
the generations of Union Carbide—could absorb the tragedy and march on.

A year after the disaster, Anderson was declaring that his own trauma had
passed and that the situation had become "manageable" and "nonemotion-
al."[70] In a letter to *Business Week* where he reiterated Union Carbide's never-
proven theory that the disaster had been a result of sabotage, he declared,
"Where Bhopal is now a symbol of disaster, there is the opportunity to make
it a symbol of recovery and hope."[71] The saboteur, said to be a disgruntled
employee, became not only a legal weapon, but also the scapegoat for emotions dangerous to the company, such as anger and disloyalty. Union Carbide first blamed a Sikh employee who had posted bail for UCIL's managing director, and then Mohan Lal Verma, a trainee in the MIC unit who held
a postgraduate math degree and was said to be resentful of his lack of progress
and possible transfer to another unit.[72] UCIL workers were also blamed for
covering up the sabotage in solidarity. Similarly, U.S. workers were faced with
transfers and layoffs in the company's post-Bhopal restructuring and were
said to be "depressed and angry."[73]

Replacing such potentially damaging emotions with "heart" and "hope"
in public statements about Union Carbide's strength and viability helped
contain them within the corporate project. The transformation of emotions
into stratified symbols of company loyalty fulfilled the CEO's expected managerial role. The national media showed workers rallying at Union Carbide's
plant in West Virginia and reported that these workers were "more loyal than
before," snapping up Carbide caps, "Kiss a Carbider" T-shirts, and "I Am a
Carbide Supporter" bumper stickers.[74] This offset the news of angry Indian
workers protesting the official closing of the Bhopal plant, staging a mock
funeral for an Anderson effigy, and chanting, "Killer Carbide."[75] Already divesting itself of its Bhopal liabilities, Union Carbide focused on maintaining faith in its U.S. operations.

Like the lifeboat, the "leaner" Union Carbide could present itself as a more
perfectly governed institution that required absolute loyalty for survival. As
the "Chemco" researchers suggested, "Crises enable the corporation to ex-
ert tighter hierarchical control" over its employees.[76] Although they were
referring to implementing safety standards through internal reports and
memos, crisis management extends to employee training, drills, and reedu-

cation in company loyalty and public relations. The routines of workers are altered not only by regulation of the body through new surveillance proce-dures, operations, and chronobiological monitoring,[77] but by the demands of postdisaster information control. As one management publication sug-gested, evoking the *Exxon Valdez* and Bhopal and warning of "malcontents," "Every employee in the entire organization is a reputation manager, from the top down."[78] Another suggested that after a crisis, employees should be sub-jected to daily briefings about what the company is doing to "master the sit-uation," equipping them to "handle questions from their friends and neigh-bors."[79] Former Lockheed Martin CEO Norman Augustine warned that in a crisis employees "should not be left to ferret out information from the pub-lic media," and extolled the action of a supermarket accused of selling bad meat: putting employees in new uniforms.[80]

The thrust of Union Carbide's internal public relations campaign was the repetition of "attitudinal symbols," a kind of recent public discourse that Leon Mayhew argues becomes detached from social groups and avoids ex-change or debate.[81] Using the infamous case of George Bush's Willie Hor-ton ads, Mayhew explains that attitudinal symbols, dissociated from any real issues, cannot be "redeemed" for any explanation, nor can they be appealed or rebutted. Because the institutional response to disaster is to create strict frames of reference that deny and evade its influence, the deployment of at-titudinal symbols absorbs the violence in their repetition. The organization is reconstructed through a language of new productivity and a flow of con-sumer goods stamped with the infinitely reproductive logo. The new uni-forms, like the jumpsuits of Biosphere 2, stand for sequence and accumula-tion in the face of disaster.

These strategies extended beyond Union Carbide's image reconstruction to the entire chemical industry. As a response to the Bhopal disaster, the Chemical Manufacturers Association (CMA) instigated "Responsible Care," a voluntary safety program first developed in Canada, an attempt to reassure insurance companies and avoid external auditing by government regulators.[82] As a prerequisite for CMA membership, chemical companies must agree to abide by management principles that include community outreach, safe stor-age and handling of chemicals, and prompt reporting of hazards. All of these are based on a self-auditing system, currently without official third-party verification. Introduced on Earth Day in 1990, Responsible Care's logo con-sists of two blue hands reaching up, almost as if in prayer, yet opened to re-lease a fruit or seedlike model of a molecule, as if in offering. This image, at once spiritualizing and naturalizing the hands as if they are growing from the ground, was designed to gain public trust with a new attitude of humility, replacing the chemical industry's culture of closed arrogance and elitism with a folksy organic community presence. In a Union Carbide advertisement during the "Green Revolution," it promoted itself as a "hand in things to come," a godlike white hand pouring pesticides from a test tube onto a field being tilled near the Ganges, its gleaming chemical plant magically floating on the waters in the background. After Bhopal, that advertisement appeared in a book exposing Union Carbide's bad environmental record, accompanied an article on the disaster in the radical environmental magazine *Fifth Estate*, and appeared on an Earth First! Web site as a menacing image of contamination and poisoning.[83]

This transformation of discourse I will call the "Bhopal effect": not merely the instigation of a new safety culture within the corporation, but the tone of disaster that persistently destabilizes institutional containment efforts. The hands of Responsible Care, despite their offerings to an imagined community in a circle of consensus, still hold this menace, and these public relations efforts have not been deemed particularly successful in winning over public approval, even as reported in the industry press.[84] Memories of the Bhopal disaster, reinforced by public knowledge of frequent fires, explosions, and leaks in chemical plants that have come after Bhopal, prevent chemical industries from securing trust and legitimating themselves through a control of postdisaster discourse. Narratives of corporate identity are bifurcated by disaster, building a public relations campaign to avoid culpability and engineer consent while being forced to produce another flow of documents that reveal the gaps in that campaign. Most treatments of corporate image reconstruction see only the carefully managed public relations campaign and therefore focus on the narrow success or failure of managed image.[85] In the same vein, critics of corporate greenwashing offer a view of a static institution behind a seamless veil of deception rather than one that is also internally vulnerable to disaster (as even industry analysts suggest), a gap that may be exploitable in the interests of worker and community safety.

On the national level, Bhopal provided a symbolic basis for advocating new regulations for the chemical industry. Under lobbying from the Sierra Club, Clean Water Action, and Ralph Nader's Public Interest Research Group, who were taking advantage of new public awareness of the industry after Bhopal, the 1986 Emergency Planning and Community Right-to-Know Act (SARA Title III) was passed, including a requirement that chemical companies publish their toxic releases. Located on the EPA Web page, the Toxic Release Inventory (TRI), organized by state, is easy to access and use and presented without any "spin," an effective tool for organizing against local polluting industries. The TRI is by no means perfect. It covers only endstream emissions rather than including internal emissions, allows companies to manipulate data through the self-auditing system, ignores dangerous releases that fall below official threshold levels, doesn't require all polluting industries (such as agribusiness) to report, and doesn't cover onsite storage of toxins, the very problem at Bhopal.[86] The EPA's budget doesn't allow it to enforce its regulations sufficiently or implement monitoring programs. The TRI covers only chemical emissions from U.S. plants, making it difficult to monitor U.S.-based industries' international performance as they site dangerous technologies overseas.

However, it has forced some exposure of the normal toxic practices behind the corporate veil, an openness many have argued is necessary to the safety of workers and surrounding communities. TRI statistics leak through the gaps of a traditionally closed, protective discourse, undermining the corporation's strict control of information and identity, which it maintains through public relations and under the guise of protecting trade secrets. It is optimistic to assume that residents highly invested in local industries and living within established contexts for discussing this relationship will use TRI data for action. But the inventory allows a new range of interpretive possibilities and rearrangements of local social contracts. The TRI represents a series of everyday disasters, large and small, the intimacy and relentless ordinariness of rupture in the systems that surround us.

Safety Street

Because of the national media's short-lived and short-sighted coverage, the Bhopal disaster might have been forgotten in the United States if not for the attention leveled at a similar MIC-producing plant in Institute, West Virginia. As when the news from Hiroshima made people imagine concentric waves of blast and fire and shock emanating from U.S. cities, the national media's immediate response to Bhopal was not "How can we help the victims?" but "Can it happen here?"[87] Could death's scythe sweep sleeping towns in the nation's chemical alleys and valleys? Especially for the residents of West Virginia's Kanawha Valley, Bhopal carried the light of apocalyptic revelation, the exposure of sudden death that had been hiding behind the gates of Union Carbide and other resident chemical companies.

In his proposal for an "ethics of the future," Hans Jonas advocates a "heuristics of fear," the extrapolation of worst-case scenarios that would provide a corrective to the rapid, destructive pace of technological development.[88] Jonas suggests that engineering requires short-range predictions about how things will work, but that engineers fail to see the full range of effects and the potential totality of destruction. Therefore, "we" have a duty to be open to fear, and to imagine the worst, carrying out "well informed thought experiments" and acting accordingly.[89] Jonas suggests that this is a shared scientific and philosophical project, but mentions science fiction as an appropriate venue for such prognostications. Imaginations of disaster, or thought experiments, do not carry the same weight as a disaster such as Bhopal, embodied and grounded in suffering on a massive scale. Knowledge of the suffering at Bhopal, which even the national news could not mitigate completely, has provided a lasting shape to fear, especially in Institute, with its historical links to Bhopal. The Bhopal effect brought the Institute plant to the surface of public attention and presented an acute image of a potentially disastrous technology, at odds with the opaque rationality of an industry that had receded into the routines of everyday life. The Kanawha Valley was already well known to environmentalists as Cancer Valley because of its high disease and mortality rates, but the slow leaking of industry-contested statistics about carcinogens was not nearly as gripping as the shocking footage of the dead in India.

The Institute and Bhopal plants were held together by flows of discourse and technologies, engineering transactions and mediated affiliations, and chemical leaks that harmed their residents. Union Carbide had reason to worry about perceptions of its West Virginia plant, which it closed for an overhaul of safety systems within a few days of the disaster in India. Union Carbide initially stated that the two facilities were "nearly identical."[90] The design package for Bhopal's MIC production facility, including flow diagrams and specifications, was based on the design and performance of the Institute plant. There was a flow of expertise from West Virginia to Bhopal during the plant's startup in 1980, and the U.S. team leader who oversaw its initial operations stayed on as works manager for two years. Both facilities stored hazardous amounts of MIC. As over a hundred news organizations inundated Institute looking for a related story in the aftermath of the disaster in India, the town's community relations, industries, and identity were transformed as it became Bhopal's analogue. However, because Bhopal, like Hiroshima, was an event beyond prediction, beyond the bounds of representation, Institute was a controllable environment where the End could be evoked but never fulfilled.

For a small number of Institute residents, mostly academics at West Virginia State College (WVSC), the news from Bhopal provoked feelings of sympathy and solidarity and a movement to end hazardous chemical production in the valley. At a town meeting in the neighboring Dunbar, six days after the Bhopal leak, Edwin Hoffman, a professor of history at WVSC, pointed out similarities between Bhopal's victims and Institute's predominantly African-

American population and urged the Indian people to sue and imprison Carbide officials. He suggested, "We are in equal jeopardy and have been just the lucky ones—there but for the grace of God go I."[91] Implications of Union Carbide's environmental racism, both abroad and at home, made parallels between Bhopal and Institute even stronger, beyond technical operations and design. The Institute plant is sited close to an African-American neighborhood, and the town hosts WVSC, which began in 1891 as the West Virginia Colored Institute. The area has a long history of civil rights and union activism, and the desegregation of Charleston is still well remembered.[92]

Hoffman organized a small group, called People Concerned about MIC, who met at the college to express their concerns, especially over the impending restart of MIC production at the plant. They drew up a fact sheet on MIC that showed India and Kanawha County side by side, equal in scale, with Bhopal and Institute marked, suggesting a firm kinship and equal risk of destruction. In a letter to the people of Bhopal, from the "shadow of the Union Carbide chemical plant," they offered their hand in a bond of "common concern for safety and health."[93] One of the members, art professor Paul Nuchims, sent a satirical letter to Warren Anderson inviting him to stay at his house, 300 yards downwind, so that Anderson could share his "views of life and corporate responsibility."[94] Nuchims explained that his kids had gone

facts about MIC

India   Kanawha County
West Virginia

AT LEAST 2,500 PEOPLE DIED ON THE NIGHT OF DECEMBER 3, 1984, AS A RESULT OF A LEAK OF MIC GAS FROM A UNION CARBIDE PLANT IN BHOPAL. IT IS POSSIBLE THAT 5,000-15,000 DIED. 200,000 WERE AFFECTED BY THE TOXIC GAS, AND 60,000 MAY SUFFER PERMANENT EFFECTS.

WHAT HAPPENED IN BHOPAL?
WHAT IS MIC?
HOW IS MIC USED?
WHAT SHOULD BE DONE?

The first brochure from People Concerned about MIC, showing India and West Virginia on the same scale. (Courtesy of People Concerned about MIC)

to Carbide summer camp and that Institute was lovely in spring. Anderson pleaded other commitments. A mordant wit often emerged in the group's statements and newsletters, which often included political cartoons.

Loyal to its industries, the community mostly dismissed the group's efforts until an incident at the Union Carbide plant raised the apparitions of Bhopal. When the plant was put back into production, with new safety systems, eight months after the Bhopal disaster, it leaked a cloud of toxins mostly consisting of aldicarb oxime and the carcinogen methylene chloride, injuring 145 people and hospitalizing 31. As Charles Perrow had warned, new safety devices actually made the system more vulnerable: The aldicarb oxime production unit had been installed to process MIC into a "less toxic" substance as part of the post-Bhopal renovation.[95] Although it typically ignored (and still ignores) countless "minor" incidents such as this, the national media had a highly charged preexisting framework for heightening the sensation of the leak: "A choking cloud of gas derived from the chemical blamed for the deaths of about 2,500 people in India last year leaked Sunday from a Union Carbide plant . . . and rolled 'like a fog' through four towns."[96] Initial reports suggested that MIC residue was present in the release, but company spokespersons stated that the gas was simply brake fluid, attempting to contain it within familiar terms.[97]

As in depictions of Bhopal, the milky cloud of chemicals wafted over the town, this time in daylight, smelling of cat litter, gasoline, or rotten eggs rather than chilies. As at Bhopal, the siren failed to sound. People were stricken suddenly at their everyday activities. But these national media stories, while making comparisons to Bhopal, suggested economic and technological differences in ways of life. Residents were stricken on open golf courses and tennis courts rather than in their crowded slum huts. Some fled in their cars, a means of evacuation not readily available in Bhopal. News footage showed the injured being carried on stretchers, with IVs in their arms, rather than lying dead in the streets, suggesting that public health services were prepared. No one was killed, few were injured, people were used to it, it wasn't worth moving out of town over, the plant wasn't closing down, the bargain could still be made. Because the disaster fell short of Bhopal, the Hiroshima of the chemical industry, it was only a warning that proved the survivability of the local system.

But Institute now had its own disaster victims, and fears that they could be "Bhopaled" took on a greater intensity. People Concerned about MIC sponsored a town meeting, inviting Union Carbide officials to speak. *Chemical Valley*, a 1991 documentary by Appalachian filmmakers Mimi Pickering and Anne Lewis, includes footage of this well-attended meeting, in which local African-American environmental activists, mostly women, shouted down Union Carbide officials and critiqued company statements about chemical releases and evacuation methods. Union Carbide president Robert Kennedy, who later replaced Anderson as CEO, was interrupted by an angry worker when he tried to tell a story about his aggressive dog. *Chemical Valley* begins with news footage from Bhopal before it moves to the angry meeting, suggesting that if Bhopal's victims seemed passive in their stunned grief, they found a public voice for protest in the United States through the residents of Institute.

In practical terms, the aldicarb leak's radicalizing of many Institute citizens, including plant workers, led to further demands for change. Union Carbide sold its Institute plant to Rhone Poulenc two years after Bhopal, divesting itself of what Robert Kennedy called the "enormous burden of the terrible publicity," but kept its South Charleston plant.[98] With its new allegiances of residents and workers, People Concerned about MIC, which had initially hoped to ban dangerous (but not all) chemical production from the valley, turned its attention to education, legislation, and community safety and to preserving the memory of Bhopal.[99] The group took on a civil defense role, fighting for a new evacuation route from the college in case of disaster and critiquing safety drills. In 1992, its chair, Pam Nixon, who had been a victim of the aldicarb leak, wrote a letter to the Local Emergency Planning Committee asking for worst-case scenarios from the area's chemical companies. They cooperated because they would eventually be required to produce worst-case scenarios under SARA Title III anyway.

Nationally, spokespersons for the chemical industry worried that worst-case scenarios would be a public relations nightmare.[100] Managers feared losing legitimacy outside the corporate environment as the scenarios were presented in public forums: "In these settings, executives lack the control and authority that they command within their organizations."[101] Evoking the specters of Bhopal, an Atochem plant manager said of presenting scenarios to the community, "The question is how you communicate in a manner that is real and truthful, but doesn't keep them awake at night thinking they'll be gassed in their beds."[102] The imagination of disaster, forced by mandatory production of worst-case scenarios, takes its material from prior representations of disaster as well as from technical knowledge of design and engineering. Chemical companies see their scenarios as proving that the disaster will always fall short of the very worst.

In 1994, the Kanawha Valley chemical companies sponsored a $60,000 event at the Charleston Civic Center and Town Center Mall to deliver their worst-case scenarios and presumably soften their reception. Through presen-

tations and information booths, chemical company spokespersons attempted to define this release of information as a showcase for safety systems already in place according to existing standards set by "all the stakeholders."[103]
People Concerned about MIC also had a booth at which they showed a film
about Bhopal, making sure that these images were still in the minds of residents. The event was called Safety Street, suggesting that the chemical industry
was like the traffic system. Typically, in discussions of risk, driving a car is seen
as a chosen rather than inflicted risk. In their descriptions of the toxic chemicals featured in the worst-case scenarios, most brochures were careful to point
out their household uses in synthetic leathers, paints, detergents, pharmaceu-
ticals, toothpaste, bowling balls, furniture, and the like. Thus, carcinogens,
mutagens, and severe irritants, some with not completely understood health
effects, were domesticated and shown to be consumer's choice. In a rather
typical exercise of engineering consent by making danger seem attractive,
Safety Street promoted a vision of a community choosing and managing
chemical risks together, of being "good neighbors."

What is striking about the worst-case scenarios presented by the eight
chemical companies—Monsanto, FMC, Rhone Poulenc, ARCO, Union Carbide, Olin, Du Pont, and Oxychem—is that they resonate with cold war civil defense images and tactics.[104] Each scenario describes a toxic chemical
plume that spreads from the plant in concentric ellipses of danger, the inner
circle being the highest concentration of the chemical that will produce the
greatest health effects. Civil defense diagrams of nuclear blast effects used the
same imagery of concentric circles, suggesting a rationality, predictability, and
limitation to the unfolding catastrophe. At Safety Street, participants could
interact with these scenarios via CD-ROM.[105] Joyce Carol Oates, in her novel
You Must Remember This (1987), describes the psychology of the old civil
defense maps through the eyes of Enid Stevick, a schoolgirl living in a work-
ing-class neighborhood: "If the bomb was dropped it would be dropped on
the factories by the lakeshore, which meant some of the East Side was within the 10-MILE RADIUS circle, though the De Witt Clinton Junior High
School itself and the stretch of East Clinton where the Stevicks lived was inside the 20-MILE RADIUS circle. That was good, they said. That wasn't too
bad, they said."[106] Similarly, the rhetoric that accompanied these scenarios
reassured residents that wherever they stood, it wouldn't be too bad. The
chemical companies themselves confessed that their worst-case scenarios
weren't the "extreme worst" because they covered only single-tank incidents,
not accidents involving entire processes.[107]

Each chemical company's brochure also described what it was doing to
prevent the disaster and outlined its emergency response. According to their

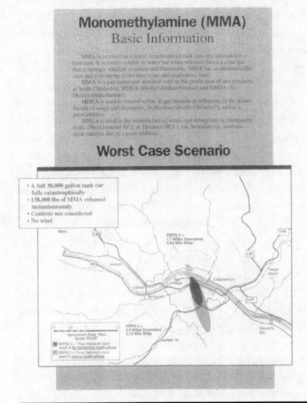

Worst-case scenario from the *Safety Street* pamphlet distributed by Union Carbide. (Courtesy of Union Carbide Corp.)

literature, the plants have regular drills, emergency response personnel, and an emergency warning system, including the Emergency Broadcast System and factory sirens. (Union Carbide's alarm did not go off when the aldicarb leaked.) Citizens are expected to be involved in emergency planning, cognizant of the warning system, and ready to "shelter-in-place" rather than run for their cars. Like the old fallout shelters of the 1950s, rooms of public buildings, like the college, are now designated shelters from chemical releases, equipped with radios and tape for sealing them off. Building wardens have been assigned to supervise the hunkering down in an orderly manner. And just as in the old civil defense drills, students at the college joke about the futility of the exercise: "When it happens I'll just bend over and kiss my butt goodbye."[108]

Just as cold war civil defense drills both provoked apocalyptic fears and reinvented the Bomb as a disaster that might be survived, these worst-case scenarios suggest that the solution to chemical dangers is not to eradicate them, but to rationalize them within preexisting frames of reference. An educational video to train residents in emergency procedures included a firefighter who explained, "An emergency plan is no different than a game plan for the super bowl, if we don't practice the plan it won't work very well."[109] He also advised preparing a home emergency kit including a radio, flashlight, batteries, first aid supplies, candles, matches, bottled water, and duct tape. (I remember my mother putting these same things in a crawl space during the Cuban Missile Crisis, with the exception of the duct tape.) James Brandt, the American Red Cross's Disaster Response Coordinator, said, "There is really nothing mysterious about a community pulling together, it simply means getting to know your community and then 'joining up' as a participating citizen of the community. A fenced-in chemical plant can look pretty threatening to most people, but if you become familiar with the folks who run the plant, respond to emergencies, plan for the response and oversee the planning process you'll realize they're normal, caring human beings."[110] He went on to point out that pesticides, lead, asbestos, organic solvents, and other toxic materials can be found in the average home. Once again, this normalizing of disaster, designed to soothe the fears of a threatened populace, accentuated the dangers of everyday life. According to the contradictory rhetoric of emergency planning, the domestic sphere is fraught with the same dangers as a chemical plant, but offers the only shelter from them.

A survey to assess the impact of Safety Street on the community found that little had changed in people's attitudes toward chemical companies, which are mostly favorable, or their knowledge of safety procedures.[111] People Concerned about MIC continues to gather at the college to discuss the latest leaks and fires from its good neighbors, the chemical companies. After a fire and release of toluene vapor at Rhone Poulenc's pesticide plant in 1996 that forced thousands to shelter-in-place for an hour, a former Union Carbide pipefitter spoke up at the group's town meeting: "Bhopal is going to pale in comparison to what's going to happen here."[112] People Concerned about MIC's chairperson, Pam Nixon, who gained national recognition when she forced the release of worst-case scenarios, continues to demand independent safety audits and has refuted industry claims that it is reducing storage of hazardous chemicals and improving its safety record. In 1995, when a spokesperson for Union Carbide denied that a leak of dimethylethanolamine had escaped factory grounds, Nixon retorted, "Stop saying it did not get into the community. It did get by the fenceline because you don't have a bubble over your plant."[113]

In that year, Nixon told the local press that she had been diagnosed with a rare immune disorder. She described it thus: "The body looks at itself as a virus and tries to kill itself."[114] Her announcement coincided with her receipt of a Mother Jones award, given by the West Virginia Environmental Council. At the ceremony in the capitol rotunda, she had to be helped to the podium. Her illness, pemphigus vulgaris, was immediately politicized in the local press. Remembering her injury from Union Carbide's 1985 aldicarb accident, Nixon cited a study by the National Institute for Environmental Health Studies that listed toxic chemicals as a possible cause. Union Carbide spokesman Thad Epps, holding no evidence, countered that Nixon's work in a medical lab might have caused it.[115] Nixon's body became a site for local media analysis of the possibilities of chemical exposure, playing out along the lines of inseparable physiological and psychological traumas: "The disease has eaten away Nixon's skin, leaving much of her body covered at times with blisters. Steroids prescribed as a treatment brought on a mix of bloating, weakness in muscles, anxiety attacks and tremors. Nixon takes valium to control the anxiety attacks."[116] With the authority that such suffering holds in the culture of simulation, the disaster survivors of Bhopal and Institute demand a recognition of the realities of pain inflicted by technological systems. They demand a public accounting.

Conclusion

From 1990 to 1997, I asked my university students what the most significant media event of their childhood had been, and almost without exception they said the *Challenger* space shuttle disaster. They often described their teachers leading them down to special classrooms to view not the launch, which had already taken place, but the explosion. From my own childhood, I remember the eerie footage from the *Apollo* moonwalk and the pride in U.S. technological superiority that came with it; these students, who grew up in the 1980s, claimed their defining moment as national spectators in the failure of a national project. The space shuttle disaster demonstrated that highly touted technological systems, with all their investments of political faith, were crashing.

During the years of large systems building, stemming from Fordism and Taylorism and refined in the Manhattan Project, inhabitants of industrial societies in transit to the information age became increasingly dependent on air and road networks, huge weapon programs, public utilities, sophisticated medical technologies, synthetic consumer goods, agribusiness, and electronic, digital, and satellite communication. Despite occasional disasters, the ills of such systems seemed to come from their mishandling and political misuse, rather than basic flaws in their construction and ordinary operation, in their cultural embedding and increasing normality. Fears of technological systems were projected mostly onto nuclear weapons and omnicidal scenarios cooked up by a secretive nuclear priesthood, until the events I have described in this book became cultural spectacles in which more tacitly accepted systems suddenly seemed extraordinarily dangerous. Each represented a certain venue: aeronautics, medicine, nuclear energy, oil extraction, chemical production. Although more than one oil spill occurred in the 1980s, only one reached the level of spectacle, as though two oil spills would have been redundant as analytical sites for questioning technological progress. Chernobyl so surpassed Three Mile Island in its political and environmental implications that it replaced it as the representative nearly worst-case scenario for nuclear energy production. Selected disasters became public and private forums for questioning fundamental modes of a highly technological existence.

It was not surprising that institutions would rush to rebuild and launch a defense of their machines, methods, and procedures, in which their entire identities were invested. In the previous chapters, I have identified certain shared strategies used by institutions and their consultants to restabilize their identities in crisis. First of all, they reacted by attempting to contain and scale down unpredictable systems that were generating catastrophic results. Their public rhetoric emphasized restoring, stabilizing, refining, and improving technological systems for total safety, despite all evidence that they could not be made completely safe. Because safety is more conceivable in a limited environment, where near total surveillance and a minute accounting are possible, institutions imagined lifeboats, postdisaster models of themselves that incorporated tighter monitoring, a shielding against perceived external threats, greater control over employees, and increased internal documentation. These efforts were supported by academic consultants in sociology, communication, and organizational management, using their own system models, leading to new areas of specialization within disciplines. Systems of discourse became the prescription for technological ills, and dependency on technological systems was justified in the name of safety. Second, crisis communications, aimed at controlling public perception of corporations and other institutional entities shaken by disaster, became a lucrative public relations specialty. Public relations consultants stripped down the language with which government and corporate organizations communicated their images to the public, perceived as a hostile political environment. Public monuments, encasing the remains of disaster in an alternative, "safe" structure for tourism, reassured visitors that danger was contained and safely buried.

The national press, through which disasters became large-scale meaning-generating machines, both inflated them with nuclear and apocalyptic images and attempted to contain them in the language of crisis management. The voice of the press, conveyed through technological means, was in itself a reassurance that massive technological systems were still functioning and in order. Because of the formulaic standards of news writing, stories of disaster featured certain actors who performed new relationships with technologies or reconfirmed old ones, based on conventional expectations of gender, race, ethnicity, and class. At the same time that it announced high-tech anxiety, the national media generally supported versions of disaster control put forth by corporations and government agencies and promoted the preservation of these entities.

The cultural imagination of disaster in the United States demands that catastrophic events not only be survived, but also be productive. Thus, resignation and fatalism among survivors were often examined and critiqued in the media, providing broader explorations of bargains with blanketing technosystems. Corporations congratulated themselves for forming new productive liaisons with surrounding communities. Scientific emissaries and other tourists with expert knowledge visited disaster zones to demonstrate that the worst could be braved and survived. A magical fertility was associated with disaster in some popular scientific accounts, as if it had stimulated new growth and diversity. And the fertility of discourse itself stood against the ultimate silence of utter catastrophe.

Despite very powerful interests attempting to control public discourse, none of the disasters I have described could be easily disciplined within single, self-interested narratives, even if repeated over and over in various publications. Angry survivors of disaster found various means to enter the debates, and often just their physical presence, giving visible evidence of emotional and physical pain, was enough to draw media attention. The body with its vulnerabilities could not be completely absorbed into institutional and media discourses that emphasized the viability and safety of technological systems. Nevertheless, survivors found themselves negotiating new understandings of their own bodies and environments, learning the language of science and medicine and preserving prior cultural knowledges to create hybrid means of survival. To gain a voice in a spectacle that would rather have kept them silent in the interests of mythical progress, survivors used alternative modes of discourse, such as publicly reading roll calls of the dead and incorporating their names into public art works. Communities with common experiences of disaster sometimes reached across cultural divides in solidarity.

Whereas the disaster spectacles of the 1980s provided a forum for examining the accidental dangers of the complex, externally managed technological systems we now inhabit, the spectacles of the 1990s have taken a new turn toward intentional acts of violence involving these systems. The idea that terrorists can transform familiar, useful artifacts into weapons that penetrate and destroy has been played over and over in the bombings of the World Trade Center, the Oklahoma City federal building, the Olympic Games in Atlanta, U.S. embassy buildings in Africa, and the offices and households of the Unabomber's victims. Watches, knapsacks, cassette players, contact lens cases, and suitcases have become the new "lethal technologies" remaking the ordinary world into a place of apocalyptic violence. Thus, uncertainties about the potential dangers inherent in technological systems have shifted to external, pathological agents who invade, disrupt, and destroy passive structures. The exception is the current attention to the Y2K bug, which some fear will crash computers and all the systems they control in the year 2000 when internal timeclocks cannot handle the change of millennium. As of this writing, the Y2K bug unfolds in the realm of fantasy.

Yet despite the new silence surrounding them, normal accidents continued to happen in alarming numbers throughout the 1990s. Trains derailed, pipes leaked, and ships spilled their rotten cargo. In Guadalajara, a leaking Pemex oil pipeline exploded, killing 189 people. A heavy rain caused 35 million gallons of animal waste to pour into a lagoon in North Carolina, killing 10 million fish. In Guyana, a Canadian-owned gold mine suffered a dam breach, releasing a continuous flow of cyanide into the Omai Creek, completely destroying its aquatic life. Residents were forced to drink water brought in by government officials and the mining company.

A truck overturned on the Baltimore Beltway, spilling hundreds of gallons of paint thinner and injuring ten. A Southern Pacific train derailed, spilling metham sodium into the Sacramento River, where it killed all fish for 42 miles. Local residents complained of headaches and miscarriages, and clean-up workers developed dermatitis. Near Duluth, Minnesota, a Burlington Northern train derailed into the Nemadji River, spilling benzene and causing the evacuation of over 22,000 people.

In Rotterdam, Netherlands, 100 tons of hydrochloric acid were released from a Consolidated Metallurgical Industries storage facility. A recycling company caught fire in Ontario, releasing a smoke plume filled with heavy metals, benzene, and polychlorinated biphenyls over the town of Hamilton. A valve burst at a General Chemical plant in Richmond, California, sending sulfuric acid into the air, injuring 24,000 people. An explosion at Union Carbide's Seadrift, Texas, plant killed one worker and injured thirty-two others. An explosion at Arco's Channelview, Texas, plant killed seventeen workers and flattened an area the size of a city block. At Shell's Belpre, Ohio, plant an explosion killed three workers and released ethylene dibromide, a carcinogen and mutagen, into the Ohio River. The Institute MIC factory once again suffered an accident, an explosion that killed one worker and made residents very nervous.

An American Trader tanker ran aground, spilling 400,000 gallons of oil that damaged 15 miles of coastline in southern California, severely damaging the pelican population. The tanker *Julie N* spilled 170,000 gallons of oil into Portland Harbor, polluting the Stroudwater Marsh. The *Sea Empress* collided with rocks in the channel off Pembrokeshire, England, losing 72,000 tons of oil and killing off the northern populations of the cushion star. An engine on the *Braer* failed, causing it to crash and spill 80,000 tons of oil off the Shetland Islands. A Russian tanker, *Nadohka,* split in two and polluted the coasts of nine Japanese prefectures. A car ferry, the *Moby Prince,* and an oil tanker, the *Agip Abruzzo,* collided in the Mediterranean, spilling oil and killing 149 people. Tankers caught fire off the coasts of Genoa and Angola. A tanker sank near Cape Town, killing thirty-nine and dispensing 2,500 tons of oil into the rich seas.

Eight times the amount of oil spilled by the *Exxon Valdez* flowed from a damaged Gulf Canada pipeline, destroying a large area of Russia's fragile Arctic tundra and damaging the livelihood of the Komi people. The company denied that it had anything to do with the pipeline's operation. In Ecuador, 275,000 gallons of oil spilled into the richly diverse Oriente region, causing the Napo River to run black. It was by no means the first oil disaster in that region. Texaco has been a prime player in the oil industry there, along with Maxus, Occidental, Orix, and Arco. In 1997, angry Native Ecuadorans, members of the Ashuars, tired of fruitless legal wrangling, took two foreign oil industry workers hostage, demanding a $2-million ransom.

Survivors of disasters continued their struggles for compensation, restitution, and environmental restoration. When Exxon filed suit against the U.S. Attorney General, the Justice Department, and the Transportation Department to bring the ship *SeaRiver Mediterranean* (formerly the *Exxon Valdez*) into Alaskan waters, several Native villages filed a countersuit to prevent its return, citing the ship's symbolism. In turn, Exxon claimed the ship was being unfairly victimized.

Representatives of the Chernobyl liquidators marched on Moscow to protest the budget cuts that would rob them of disability benefits. Kiev residents took to the streets in anger after a leak from the still-functioning reactor complex. Chernobyl survivors joined a protest against the Watts Barr Nuclear Power Plant in Tennessee on the anniversary of the meltdown. On Capitol Hill, Aleksandr Sirota, a 19-year-old ill from radiation effects, and Dr. Sergei Paromchik called all nuclear power plants "ticking time bombs." Two visitors from the heavily irradiated Novozybkov marched with Australian environmental activists on a pilgrimage to stop uranium mining on aboriginal lands. Chernobyl refugees went back to their home towns on the tenth anniversary, laying flowers on the graves of the dead. Others chained themselves to the fence surrounding the Forbidden Zone. At a Chernobyl commemoration event in Minsk, thousands of residents fought with police after a speech by Belarus president Aleksandr Lukashenko, who offered to repopulate the contaminated lands.

In Bhopal, at the fifth anniversary, 800 survivors from the city's gas-affected shantytowns were arrested marching toward police barricades around the Union Carbide plant. At the tenth anniversary of the chemical disaster, the widows and gas survivors took once again to the streets, chanting "Killer Carbide," burning and beating its former CEO in effigy, demanding his extradition, holding a die-in at a crossroads where some lay covered with earth, painting slogans on the walls of the city, and erecting stone tablets marked with the names of the dead within the walls of the old Union Carbide plant. The long fight for compensation, justice, and a voice in the legal, political, and cultural institutions governing their lives was still strong, especially among poor Muslim women who had lived through the tragedy. Indian authorities prevented thousands of environmental activists from joining the Bhopal protesters in solidarity. A few months later, 10,000 mostly female villagers from Madhya Pradesh, where Bhopal also lies, went down to the Maheshwa Dam, cut radio communications, and began a sit-in to stop its construction. Noted activist and gas survivor Abdul Jabbar Khan was detained trying to reach a rally to protest government violence against tribal people in western Madhya Pradesh. Activist T. R. Chouhan, a former worker at the Union Carbide plant, was denied his passport renewal. Another survivor, Rehana Begam, joined a ten-year commemoration in Charleston, West Virginia, where she told listeners about women who menstruate for a month and babies born covered with blisters.

The disaster changes everything, especially for those in closest proximity to it. Although some try to contain it within monuments and narratives of containment, the new ways of life it brings must always be negotiated, sometimes privately, sometimes publicly. This book, as an archive of disaster discourse from the 1980s, continues a conversation about the meaning of these events, which have not been resolved, completely silenced, or safely buried. The technological lifeworld is still a dangerous place, even in our ordinary interactions with devices, machines, and systems. The writing of disaster always stands against the utter silence of the worst that lifeworld has to offer. I began this book with two eyewitnesses of the bombing of Nagasaki, a schoolgirl and a journalist, who experienced the moment when technological apocalypse became real and visceral. Today, the survivors of Hiroshima and Nagasaki gather every year at Hiroshima's Peace Park to remember that event, to place the names of the most recent radiation deaths in the cenotaph, and to speak of the possibilities of a nonviolent world.

Notes

Introduction

1. Hayashi Kyoko, "Ritual of Death," in *Nuke-Rebuke: Writers and Artists against Nuclear Energy and Weapons,* ed. Morty Sklar (Iowa City: Spirit That Moves Us, 1984), 30.

2. William L. Laurence, *Men and Atoms* (New York: Simon & Schuster, 1962), 155.

3. Ibid., 159.

4. Nuclear criticism is a form of literary criticism that, for the most part, attempts to examine texts for signs of a sometimes hidden "nuclear referent" and to impose a new ethical basis for writing and thinking about the ultimate disaster. As William Scheick explains, nuclear criticism "implicitly seeks to discover and uncover, by whatever act of 'intuitive rationality' necessary, some 'nuclear' features hidden within the human unconscious" and "urges humanity to reinterpret its communal memory in the context of 'living' with the threat of the nuclear referent" ("Nuclear Criticism: An Introduction," *Papers in Language and Literature* 26 [Winter 1990]: 6–7). Such critics argue that nuclear war provides the overriding, paralyzing cultural logic of nuclear powers and that some way must be found for thinking beyond this logic, which is in fact only an apocalyptic fantasy, a simulation, an entirely textual event. Thus, language must be interrogated for nuclear thinking. This work includes compilations of nuclear doublespeak dictionaries, examinations of masculinist discourse in the military and nuclear sciences, and analyses of nuclear war's representation in popular culture, literature, and the fine arts. Nuclear critics see language not only as the shaper of nuclear culture, but as the only way out of its dilemmas.

Although my work is not as committed to a psychological examination of nuclear texts and contexts, it does glean many insights from this body of criticism. Since 1945, nuclear war has been the Western apocalypse and stands as the paradigm of disaster and survival logic. In the 1990s, as I write, the apocalypse has been shifting to the more subtle forms of viral invasion, global warming, sperm count loss from pollution hazards, and the like. But nuclear criticism is not yet outdated.

5. Jacques Derrida, "No Apocalypse, Not Now (Full Speed Ahead, Seven Missiles, Seven Missives)," trans. Catherine Porter and Philip Lewis, *Diacritics* 14 (Summer 1984): 28.

6. Ibid.

7. Peter Schwenger, *Letter Bomb: Nuclear Holocaust and the Exploding Word* (Baltimore, Md.: Johns Hopkins University Press, 1992), 149.

8. Donald Pease argues, "As the representation of anticipated total disaster, Hiroshima transfigured Cold War spectators into symbolic survivors of their everyday lives, able to encounter everyday events as the afterimages of ever-possible nuclear disaster" ("Hiroshima, the Vietnam Veteran's Memorial, and the Gulf War: Post-National Spectacles," *Surfaces* 4.206: 7–8; available online at http://tornade.ere.umontreal.ca/~guedon/Surfaces/vol4/pease.html, 18 Sept. 1998).

9. Don Ihde, *Technology and the Lifeworld: From Garden to Earth* (Bloomington: Indiana University Press, 1990), 144. For a more specific discussion of the pluricultural shape of the modern media, see his "Image Technologies and Traditional Culture," in *Postphenomenology: Essays in the Postmodern Context* (Evanston, Ill.: Northwestern University Press, 1993), 43–55.

10. The term *stakeholders* is used in organizational and public policy studies to describe business and community relations. Theoretically, stakeholders are those who have the moral and philosophical right to participate in a business's decisions, but in practice they are most often narrowly defined as legal entities. For the application of stakeholder theory to disaster studies, see Paul Shrivastava, *Bhopal: Anatomy of a Crisis* (Cambridge, Mass.: Ballinger, 1987); and Thomas Birkland, *After Disaster: Agenda Setting, Public Policy, and Focusing Events* (Washington, D.C.: Georgetown University Press, 1997).

Diane Vaughan, in her study of the *Challenger* disaster, broadens this organizational view, arguing that all people have a "frame of reference," a set of "assumptions, expectations, and experience," that shapes their worldviews. Thus, she argues, different people had different assessments, shaped by culture, of the *Challenger* disaster. However, like many crisis management specialists, she limits her investigation to internal organizational communications and decisions, applying the theory to the work groups at NASA (*The Challenger Launch Decision: Risky Technology, Culture, and Deviance at NASA* [Chicago: University of Chicago Press, 1996]). Other disaster theorists limit their investigation to crisis communications, as I discuss throughout this book. For example, William Benoit studies corporate image management after disaster in *Accounts, Excuses, and Apologies: A Theory of Image Restoration Strategies* (Albany: State University of New York Press, 1995). Still other management studies focus on organizational "deviance" or distorted group thinking that causes disaster to occur. See M. David Ermann and Richard J. Lundman, *Corporate and Governmental Deviance: Problems of Organizational Behavior in Contemporary Society* (New York: Oxford University Press, 1982); Thierry C. Pauchant and Ian I. Mitroff, *Transforming the Crisis-Prone Organization: Preventing Individual, Organizational, and Environmental Tragedies* (San Francisco: Jossey-Bass, 1992); and Ian I. Mitroff and Thierry C. Pauchant, *We're So Big and Powerful Nothing Bad Can Happen to Us: An Investigation of America's Crisis-Prone Corporations* (Secaucus, N.J.: Birch Lane, 1990).

The work on disaster communities often takes the shape of disaster planning and preparedness, but sociologist Kai Erikson's work explores the impact of disaster on changing ideas of community in *A New Species of Trouble: Explorations in Disaster, Trauma, and Community* (New York: Norton, 1994).

I am most indebted for my own methodology on cultural histories of disaster that focus on their "social drama" and the multiplicity of voices in shaping their meaning. See especially Carl Smith, *Urban Disorder and the Shape of Belief: The Great Chicago Fire, the Haymarket Bomb, and the Model Town of Pullman* (Chicago: University of Chicago Press, 1995); and Steven Biel, *Down with the Old Canoe: A Cultural History of the* Titanic *Disaster* (New York: Norton, 1996).

11. Ihde sees a historical trajectory from technologically minimalist cultures, using a few simple technologies, to technologically maximalist cultures, striving for self-enclosure in a "technological cocoon," with complex instruments, systems, and procedures. However, he is careful to point out that all cultures are technologically embedded and that minimalist cultures are as "totalizing" as maximalist ones, taking "nature into culture" in their own knowledge practices: "Insofar as there are a limited number of types of human-technology relations following from human existential structures, all cultures exemplify the full range of these relations (invariantly), although the mixes are clearly variant" (Ihde, *Technology,* 124).

12. Bruno Latour, "Pragmatogonies: A Mythical Account of How Humans and Nonhumans Swap Properties," *American Behavioral Scientist* 37 (1994): 791–808.

13. Albert Borgmann, *Technology and the Character of Contemporary Life* (Chicago: University of Chicago Press, 1984), 40–47.

14. John Hersey, *Hiroshima* (New York: Knopf, 1965), 15.

15. Ibid.

16. Ibid., 23.

17. Thomas Hughes, *American Genesis: A Century of Invention and Technological Enthusiasm, 1870–1970* (New York: Viking, 1989), 442.

18. Charles Perrow, *Normal Accidents: Living with High-Risk Technologies* (New York: Basic Books, 1984), 3–5.

19. Ibid., 5–9.

20. Don DeLillo, *White Noise* (New York: Penguin, 1985), 326.

21. Disaster psychologist John Leach identifies two strategies of long-term survival for victims after the initial shock: adaptation and consolidation. The first involves a forgetting of old behaviors and a learning of "new patterns which fit more closely the new environment." The second is characterized by an attempt to reestablish an identity that acknowledges the past while accepting the new conditions. See *Survival Psychology* (London: Macmillan, 1994), 176. I'm borrowing his terms to speak in a broader sense about cultural negotiations of disaster.

22. *Exxon Valdez* Oil Spill Trustee Council, mission statement, 30 Nov. 1993.

23. Richard Klein, "The Future of Nuclear Criticism," *Yale French Studies* 77 (1990): 80.

24. Ibid.

25. Maurice Blanchot, *The Writing of the Disaster,* trans. Ann Smock (Lincoln: University of Nebraska Press, 1986), 1.

26. Murray Edelman, *Constructing the Political Spectacle* (Chicago: University of Chicago Press, 1988), 12.

27. Ibid., 10.

28. Smith, *Urban Disorder,* 4.

29. The term was used by O. K. Werckmeister to describe the implementations of a global military network with the aim of total surveillance. Such a culture, he wrote, was "dependent on the decisionless perpetuation of its crises" (*Citadel Culture* [Chicago: University of Chicago Press, 1991], 22).

30. Recent critical studies of the mass media's event selection and risk coverage include Lynne Masel Walters, Lee Wilkins, and Tim Walters, eds., *Bad Tidings: Communication and Catastrophe* (Hillsdale, N.J.: Erlbaum, 1989); Eleanor Singer and Phyllis M. Endreny, *Reporting on Risk: How the Mass Media Portray Accidents, Diseases, Disasters, and Other Hazards* (New York: Russell Sage Foundation, 1993); Conrad Smith, *Media and Apocalypse: News Coverage of the Yellowstone Forest Fires,* Exxon Valdez *Oil Spill, and Loma Prieta Earthquake* (Westport, Conn.: Greenwood Press, 1992); and Joan Deppa, *The Media and Disasters: Pan Am 103* (New York: New York University Press, 1994). These build on earlier works that describe news as a social construction rather than a collation of facts: Herbert J. Gans, *Deciding What's News: A Study of* CBS Evening News, NBC Nightly News, Newsweek, *and* Time (New York: Vintage, 1980); and Gaye Tuchman, *Making News: A Study in the Construction of Reality* (New York: Free Press, 1978).

31. Schwenger, *Letter Bomb,* 25.

32. Jean Baudrillard, *The Illusion of the End,* trans. Chris Turner (Stanford, Calif.: Stanford University Press, 1994), 3.

33. Ibid., 63.

34. Ibid., 4.

35. John Fiske, in *Media Matters: Everyday Culture and Political Change* (Minneapolis: University of Minnesota Press, 1996), argues that media analyses be based in a theory of discourse, recognizing social contexts and relations in which meanings are circulated (3).

Chapter 1: Lifeboat Ethics

1. *Lifepod,* dir. Ron Silver, with Robert Loggia, Stan Shaw, and Ron Silver, Cabin Fever Entertainment, 1993; *Poseidon Adventure,* dir. Ronald Neame, with Gene Hackman, Shelley Winters, and Ernest Borgnine, Trimark, 1972; *Lifeboat,* dir. Alfred Hitchcock, with Tallulah Bankhead, William Bendix, and Walter Slezak, Trimark, 1944.

2. According to John Leach, in actual disaster circumstances, survivors are much more likely to be calm, stunned, or bewildered, with only 10 to 15 percent showing "a high degree of uncontrolled and inappropriate behavior." Furthermore, panic is a result of "a perceived time or space limit on survival such as in a building fire or the flooding of a ship or mine." In Leach, *Survival Psychology* (Basingstoke, U.K.: Macmillan, 1994), 23–25, 30–35. Ironically, tight enclosure is more likely to breed panic than to assuage it.

3. In the early 1970s, the lifeboat ethic was articulated by social biologist Garrett Hardin as an argument against immigration and foreign aid. The United States, he suggested, was a rich lifeboat with limited carrying capacity that had to conserve its resources and space against the poorer lifeboats. He also maintained that those who gave up their lifeboat seats so that others might live were "guilt addicts." In Hardin, "Living on a Lifeboat," *Bioscience* 24 (1974): 561–68. For a more complete discussion of the metaphor and its subsequent critiques, see George R. Lucas and Thomas W. Ogletree, eds., *Lifeboat Ethics: The Moral Dilemmas of World Hunger* (New York: Harper & Row, 1976).

4. Michel Foucault, *Discipline and Punish: The Birth of the Prison,* trans. Alan Sheridan (New York: Vintage, 1979), 198.

5. Jerome Clayton Glenn and George S. Robinson, *Space Trek: The Endless Migration* (Harrisburg, Pa.: Stackpole, 1978), 184.

6. Arthur C. Clarke, *The Exploration of Space* (New York: Harper & Row, 1959), 179.

7. Jerry Grey, *Beachheads in Space: A Blueprint for the Future* (New York: Macmillan, 1983), 3, 5.

8. Gregory Whitehead, "The Forensic Theatre: Memory Plays for the Post-Mortem Condition," *Performing Arts Journal* 12 (Spring 1990): 100–101.

9. Qtd. in Karen Boehler, "Lifeboat to Safer Shores: Humans Are Still the Space Program's Most Precious Cargo," *Ad Astra*, Mar. 1989: 12.

10. In 1993, a group of naval architects and marine engineers found that the *Titanic* disaster was caused by "brittle fracture" in the low-grade steel plates of the ship's hull, caused by cold water temperatures. See William J. Broad, "New Idea on *Titanic* Sinking Faults Steel as Main Culprit," *New York Times*, 16 Sept. 1993: A8.

In his book on the *Titanic* exploration, science writer Charles Pellegrino, who had begun his own writing career covering the space program, asserted that, like the *Titanic*, "the *Challenger* was killed by ice—ice where it did not belong . . . ice warnings that were ignored in order to meet a schedule," in Pellegrino, *Her Name, Titanic: The Untold Story of the Sinking and Finding of the Unsinkable Ship* (New York: McGraw-Hill, 1988), 12. William F. Buckley, who would later visit the *Titanic* wreck in the French submersible *Nautile* declared that natural human pride and fallibility sank both the *Titanic* and the *Challenger*, in "Man is the Shuttle," *National Review*, 28 Feb. 1986: 17. Popular science historian Timothy Ferris wrote that the *Titanic*, the *Challenger*, and Chernobyl proved that "blind faith in technology can kill," in "The Year the Warning Lights Flashed On," *Life*, Jan. 1987: 67. See also Steven Biel, *Down with the Old Canoe: A Cultural History of the Titanic Disaster* (New York: Norton, 1996), 223–25.

11. E. Foster-Simeon, "Picking Up the Pieces," *All Hands*, June 1986: 19.

12. National Commission on Space, *Pioneering the Space Frontier: The Report of the National Commission on Space* (New York: Bantam, 1986), 100.

13. Diane Vaughan, *The Challenger Launch Decision: Risky Technology, Culture, and Deviance at NASA* (Chicago: University of Chicago Press, 1996), 40.

14. Testimony of Roger Boisjoly, U.S. Presidential Commission on the Space Shuttle *Challenger* Accident, *Report to the President* (Washington, D.C.: The Commission, 1986), 785.

15. Elaine Scarry, *The Body in Pain* (New York: Oxford University Press, 1985), 304.

16. Ibid., 298.

17. Malcolm Gladwell, "Blowup," *New Yorker*, 22 Jan. 1996: 32.

18. William P. Rogers, "Preface," *Report to the President*, 1.

19. Ibid.

20. Roger Boisjoly, "Interview with Tony Chiu," *Life*, Mar. 1988: 220.

21. Ron Westrum, *Technologies & Society: The Shaping of People and Things* (Belmont, Calif.: Wadsworth, 1991), 259.

22. Rogers, 1.

23. Foster-Simeon, 22.

24. Storer Rowley, "NASA Debates Whether Crew Was Aware in Shuttle Plunge," *Chicago Tribune*, 25 Apr. 1996: C2.

25. Testimony of Stanley Klein, *Report to the President*, 213.

26. Testimony of P. J. Weitz, *Report to the President*, 1437.

27. Ibid.

28. Harry L. Shipman, *Space 2000: Meeting the Challenge of a New Era* (New York: Plenum, 1987), 315.

29. Ibid., 331.

30. Anastasia Toufexis, "Good Data and a Feces Crisis," *Time,* 13 May 1985: 61.

31. James A. Van Allen, "Myths and Realities of Space Flight," *Science* 232 (1986): 1075.

32. Marvin Minsky, "NASA Held Hostage: Human Safety Imposes Outlandish Constraints on the U.S. Space Program," *Ad Astra,* June 1990: 36.

33. Stephen B. Hall, ed., *The Human Role in Space: Technology, Economics and Optimization* (Park Ridge, N.J.: Noyes, 1985), v.

34. Ibid., 63.

35. Ibid., 38.

36. Shoshana Zuboff, *In the Age of the Smart Machine: The Future of Work and Power* (New York: Basic, 1984), 23.

37. Ibid., 6.

38. Ibid., 7, 414.

39. Ibid., 414.

40. Malcolm McConnell, Challenger: *A Major Malfunction* (New York: Doubleday, 1987), 94.

41. Constance Penley, *NASA/TREK* (New York: Verso, 1997), 25, 59.

42. Lenore Terr briefly describes the effects of the *Challenger* on children in *Too Scared to Cry: Psychic Trauma in Childhood* (New York: Harper & Row, 1990), 324–29. In the 1980s the psychological literature on the effects of disaster spectacles on children burgeoned, especially around the *Challenger* disaster and the nuclear film *The Day After* (dir. Nicholas Meyer, New Line Home Video, 1983). Because children are culturally perceived as more sensitive to trauma, their thoughts and memories were subjected to intense scrutiny during the decade of disaster, for evidence that the technological world was indeed damaging even to witnesses.

43. Collected by Elizabeth Radin Simons, "The NASA Joke Cycle: The Astronauts and the Teacher," *Western Folklore* 45 (1986): 269.

44. Ibid., 272.

45. Collected by Willie Smyth, "Challenger Jokes and the Humor of Disaster," *Western Folklore* 45 (1986): 244.

46. See Simons, 261–77; Smyth, 243–60; Patrick D. Morrow, "Those Sick Challenger Jokes," *Journal of Popular Culture* 20 (Spring 1987): 175–85; Elliot Oring, "Jokes and the Discourse on Disaster," *Journal of American Folklore* 100 (1987): 276–87; and Nicholas von Hoffman, "Shuttle Jokes," *New Republic,* 24 Mar. 1986: 14.

47. Diane Vaughan, "Autonomy, Interdependence, and Social Control: NASA and the Space Shuttle *Challenger,*" *Administrative Science Quarterly* 35 (1990): 232.

48. Robert Bazell, "NASA's Mid-Life Crisis," *New Republic,* 24 Mar. 1986: 12.

49. Eliot Marshall, "The Shuttle Record: Risks, Achievements," *Science* 231 (1986): 664; R. Jeffrey Smith, "Shuttle Inquiry Focuses on Weather, Rubber Seals, and Unheeded Advice," *Science* 231 (1986): 911.

50. Smith, 911.

51. Charles Perrow, "Risky Systems: The Habit of Courting Disaster," *Nation,* 11 Oct. 1986: 354.

52. McConnell, ix.

53. Ibid., x.

54. Joseph J. Trento, *Prescription for Disaster* (New York: Crown, 1987), 4.

55. McConnell, 12.

56. Dale Carter, *The Final Frontier: The Rise and Fall of the American Rocket State* (New York: Verso, 1988), 6–7.

57. Barbara S. Romzek and Melvin J. Dubnick, "Accountability in the Public Sector: Lessons from the Challenger Tragedy," *Public Administration Review* 47 (1987): 233, 230. See also Howard S. Schwartz, *Narcissistic Process and Corporate Decay: The Theory of the Organization Ideal* (New York: New York University Press, 1990), 107–26; and C. F. Larry Heimann, "Understanding the Challenger Disaster: Organizational Structure and the Design of Reliable Systems," *American Political Science Review* 87 (1993): 421–35.

58. Michael Davis, "Thinking Like an Engineer: The Place of a Code of Ethics in the Practice of a Profession," *Philosophy and Public Affairs* 20 (1991): 150–67. Diane Vaughan disputes the idea that engineers held no blame for the disaster, while managers carried the guilt of poor decision making. She argues that both were involved in flawed communications and technical decisions (*Challenger Launch Decision*, 61–62).

59. Gregory Moorhead, Richard Ference, and Chris P. Neck, "Group Decision Fiascoes Continue: Space Shuttle Challenger and a Revised Groupthink Framework," *Human Relations* 44 (1991): 539–51.

60. Vaughan, "Autonomy, Interdependence, and Social Control," 225.

61. G. Richard Holt and Anthony W. Morris, "Activity Theory and the Analysis of Organizations," *Human Organization* 52 (1993): 102.

62. Vilmos Csányi, *Evolutionary Systems and Society: A General Theory of Life, Mind, and Culture* (Durham, N.C.: Duke University Press, 1989), 15.

63. Perrow, "Risky Systems," 354.

64. Jesco von Puttkamer, "Introduction," in Philip R. Harris, *Living and Working in Space: Human Behavior, Culture and Organization* (New York: Ellis Horwood, 1992): 9.

65. Ibid.

66. This work stems from Ilya Prigogine and Isabelle Stengers's hypothesis that chaotic systems may take up energy and begin to manifest orderly behavior. See Prigogine and Stengers, *Order Out of Chaos: Man's New Dialogue with Nature* (New York: Bantam, 1984).

67. Von Puttkamer, in Harris, 17–18.

68. Ibid., 22.

69. Jim Robbins, "Biosphere II: Our Western Home in Outer Space," *American West*, Aug. 1987: 42.

70. Charles D. Walker, "Why We Must Sail On," International Space Year special insert, *Ad Astra*, Jan.–Feb. 1992: 7.

71. Daniel Bell predicted that the industrial labor force would be replaced by workers skilled in the production and dissemination of information in *The Coming of Post-Industrial Society: A Venture in Social Forecasting* (New York: Basic Books, 1973). For a discussion of the cybernetic goals and fantasies of these knowledge workers in the late twentieth century, see Grant H. Kester, "Out of Sight Is Out of Mind: The Imaginary Space of Postindustrial Culture," *Social Text* 35 (Summer 1993): 15–32.

72. Cynthia S. Fuchs, "'Death is Irrelevant': Cyborgs, Reproduction, and the Future of Male Hysteria," *Genders* 18 (Winter 1993): 114.

73. National Commission on Space, 65.

74. Harris, 68.

75. Ibid., 95.

76. Ibid., 102. James Grier Miller began his work on general living systems theory in the 1950s at the University of Chicago's Institute of Behavioral Sciences and the University of Michigan's Mental Health Research Institute. Later, he would advocate the potential contributions of behavioral scientists to spaceology, arguing that they could plan a highly engineered human society. James Grier Miller and Jesse L. Miller, "Living Systems Applications to Space Habitation," in *Space Resources: Technological Springboards into the 21st Century,* ed. M. F. McKay (Houston: NASA Johnson Space Center, 1992); James Grier Miller, *Living Systems* (New York: McGraw-Hill, 1978); James Grier Miller, "Applications of Living Systems Theory to Life in Space," in *From Antarctica to Outer Space: Life in Isolation and Confinement,* ed. Albert A. Harrison, Yvonne A. Clearwater, and Christopher P. McKay (New York: Springer-Verlag, 1991), 177–98.

77. The term *informate* was first used by Zuboff. For a managerial view of the growing information economy, see Stephen P. Bradley, Jerry A. Hausman, and Richard L. Nolan, eds., *Globalization, Technology, and Competition: The Fusion of Computers and Telecommunications in the 1990s* (Boston: Harvard Business School Press, 1993).

78. Harris, 130.

79. National Commission on Space, 71–72.

80. David R. Criswell, "Solar System Industrialization: Implications for Interstellar Migrations," in *Interstellar Migration and the Human Experience,* ed. Ben R. Finney and Eric M. Jones (Berkeley: University of California Press, 1985), 58.

81. U.S. Office of Civil Defense, *Labor's Role in State, County, and Local Civil Defense* (Washington, D.C.: Government Printing Office, 1966), 37–41.

82. Charles E. Fritz, "Some Implications from Disaster Research for a National Shelter Program," in *Human Problems in the Utilization of Fallout Shelters,* ed. George W. Baker and John H. Rohrer (Washington, D.C.: National Academy of Sciences, 1960), 150.

83. Ibid.

84. Herman Kahn, *On Thermonuclear War* (Princeton, N.J.: Princeton University Press, 1960), 86.

85. Office of Civil Defense, *Law and Order Training for Civil Defense Emergency* (Washington, D.C.: Government Printing Office, 1965), 36–37.

86. Robert C. Suggs, *Survival Handbook* (New York: Macmillan, 1962), 190.

87. Charles Clark, "VD Control in Atom-Bombed Areas," *Journal of Social Hygiene* 37 (Jan. 1957): 1–7. Elaine Tyler May charts the connections between the nuclear family and the nuclear state, including a discussion of Clark's work, in *Homeward Bound: American Families in the Cold War Era* (New York: Basic, 1988).

88. Federal Emergency Management Agency, *Shelter Management Handbook,* in *The Nuclear Predicament,* ed. Donna Gregory (New York: Bedford, 1986), 240.

89. James W. Altman, "Laboratory Research on the Habitability of Public Fallout Shelters," in *Human Problems in the Utilization of Fallout Shelters,* 157–66.

90. See, for example, John H. Rohrer, "Interpersonal Relationships in Isolated Small Groups," in *Psychophysiological Aspects of Space Flight,* ed. Bernard E. Flaherty (New York: Columbia University Press, 1961), 263–71.

91. Martin Caidin, *The Greatest Challenge* (New York: Dutton, 1965), 290.

92. Timothy W. Luke, "Reproducing Planet Earth? The Hubris of Biosphere 2," *Ecologist* 25 (July–Aug. 1995): 161.

93. John Allen, *Biosphere 2: The Human Experiment* (New York: Penguin, 1991), 127.

94. Ibid., 21.

95. Dorion Sagan, *Biospheres: Metamorphosis of the Planet Earth* (New York: McGraw-Hill, 1990), 158.

96. Garrett Hardin, "Carrying Capacity as an Ethical Concept," in Lucas and Ogletree, 120.

97. Joseph P. Allen, "Foreword," in Abigail Alling and Mark Nelson, *Life Under Glass: The Inside Story of Biosphere 2* (Oracle, Ariz.: Biosphere Press, 1993), xi.

98. John Allen, 115. Allen was extending a concept of the noosphere (or sphere of human reason) developed by Edouard Le Roy, Pierre Teilhard de Chardin, and Vladimir Ivanovich Vernadsky in the early twentieth century. Vernadsky was most influential in his elaboration of the noosphere as a shared worldwide state of global planning and development fostered by science.

99. Alling and Nelson, 39.

100. Tim Beardsley, "Down to Earth: Biosphere 2 Tries to Get Real," *Scientific American,* Aug. 1995: 26.

Chapter 2: Sarcophagus

1. Potential water damage from fracture flows, potentially explosive buildup of gases, and migration of the radioactive contents are still unresolved issues. However, in Oct. 1997, the EPA ruled that WIPP was safe for disposal of transuranic nuclear waste and, as of this writing, the WIPP optimistically expects to begin operation by the end of the decade. See W. D. Weart et al., *Background Information Presented to the Expert Panel on Inadvertent Human Intrusion into the Waste Isolation Pilot Plant,* ed. R. V. Guzowski and M. M. Gruebel (Albuquerque, N.M.: Sandia National Laboratories, 1991), II, III.

2. Ben Finney, "One Species or a Million?" in *From Sea to Space* (Palmerston North, New Zealand: Massey University, 1992), 124.

3. OECD Nuclear Energy Agency, Working Group on the Assessment of Future Human Actions at Radioactive Waste Disposal Sites, *Future Human Actions at Disposal Sites* (Paris: OECD/NEA, 1995), 43.

4. Alan Burdick, "The Last Cold War Monument," *Harper's,* Aug. 1992: 62. See also Linda Marsa, "Bomb Shelter: Warning the Future of Our Lasting Nuclear Legacy," *Omni,* July 1993: 20; "Away!" *Mother Jones,* Sept.–Oct. 1992: 16.

5. Theodore J. Gordon et al., "Inadvertent Intrusion into WIPP: Some Potential Futures," in *Expert Judgment on Inadvertent Human Intrusion into the Waste Isolation Pilot Plant,* ed. Stephen C. Hora, Detlof von Winterfeldt, and Kathleen M. Trauth (Albuquerque, N.M.: Sandia National Laboratories, 1991), C40.

6. Dieter G. Ast et al., "Team A Report: Marking the Waste Isolation Pilot Plant for 10,000 Years," in *Expert Judgment on Markers to Deter Inadvertent Human Intrusion into the Waste Isolation Pilot Plant,* ed. Kathleen Trauth, Stephen C. Hora, and Robert V. Guzowski (Albuquerque, N.M.: Sandia National Labs, 1993), F126.

7. Burdick, 64.

8. Jon Lomberg, "Supplemental Material on the WIPP Marker," in *Expert Judgment on Markers to Deter Inadvertent Human Intrusion into the Waste Isolation Pilot Plant,* G84.

9. The accepted spelling is now *Chornobyl,* but I have maintained *Chernobyl* for consistency with most sources.

10. The first public use of *Sarcophagus* to describe the containment building appears

to have occurred during a speech by Ivan Silayev, chairman of the Soviet government's Chernobyl commission, a few weeks after the disaster. He announced the construction of a containment structure, "a sarcophagus . . . a huge container, let us say, which will enable us to secure the burial of everything that remains of the radioactive fallout of this entire accident." Qtd. in Piers Paul Read, *Ablaze: The Story of the Heroes and Victims of Chernobyl* (New York: Random House, 1993), 208.

11. Robert Peter Gale, Preface, in Vladimir Gubaryev, *Sarcophagus: A Tragedy,* trans. Michael Glenny (New York: Vintage, 1987), vii. Gale is echoing a reference to the pyramids in Gubaryev's play itself, pp. 86–87. See also Zhores Medvedev's reference to the Egyptian pyramids in *The Legacy of Chernobyl* (Oxford, U.K.: Blackwell, 1990), 20.

12. A. A. Borovai and A. R. Sich, "The Chornobyl Accident Revisited, Part II: The State of the Nuclear Fuel Located within the Chornobyl Sarcophagus," *Nuclear Safety* 36 (1995): 20.

13. Lewis Mumford, *The Myth of the Machine: The Pentagon of Power* (New York: Harcourt Brace Jovanovich, 1970), 300.

14. Ira Chernus, *Dr. Strangegod: On the Symbolic Meaning of Nuclear Weapons* (Columbia: University of South Carolina Press, 1986).

15. Helen Caldicott asserted that the nuclear arms race was the result of a "psychopathology" tied to a "tribal mentality" in *Missile Envy: The Arms Race and Nuclear War* (New York: Bantam, 1986), 229–65. Robert Lifton and Eric Markusen also promoted the idea that nuclear strategists, politicians, and arms manufacturers have a pathological mindset in *The Genocidal Mentality: Nazi Holocaust and Nuclear Threat* (New York: Basic, 1990).

16. Spencer Weart, *Nuclear Fear: A History of Images* (Cambridge, Mass.: Harvard University Press, 1988), 406.

17. For discussions of historic links between technology and the sublime, especially in a U.S. context, see Rob Wilson, "Techno-Euphoria and the Discourse of the American Sublime," *Boundary 2* 19 (Spring 1992): 205–29; David E. Nye, *American Technological Sublime* (Cambridge, Mass.: MIT, 1994); and Peter B. Hales, "The Atomic Sublime," *American Studies* 32 (Spring 1991): 5–32. Some interesting speculations on the nuclear sublime in the former Soviet Union and Japan, as well as the United States, appear in Rob Wilson and Donald Pease's interview with Oe Kenzaburo, *Boundary 2* 20 (Summer 1993): 1–23.

Commentators on nuclear language as a political tool for manipulating a passive populace are too numerous to mention. For an introduction, see Stephen Hilgartner, Richard C. Bell, and Rory O'Connor, *Nukespeak: The Selling of Nuclear Technology in America* (New York: Penguin, 1983). Donald Pease's study of callers to talk radio after the Oklahoma City bombing addresses the dismantling of cold war identities and the "sociosymbolic order" of the national security state: "Negative Interpellations: From Oklahoma City to the Trilling-Matthiessen Transmission," *Boundary 2* (Spring 1996): 1–34.

18. Jane I. Dawson, *Eco-Nationalism: Anti-Nuclear Activism and National Identity in Russia, Lithuania, and Ukraine* (Durham, N.C.: Duke University Press, 1996); Wolfgang Rüdig, *Anti-Nuclear Movements: A World Survey of Opposition to Nuclear Energy* (Essex, U.K.: Longman, 1990), 329; Miron Rezun, *Science, Technology, and Ecopolitics in the U.S.S.R.* (Westport, Conn.: Praeger, 1996); David R. Marples, *Belarus: From Soviet Rule to Nuclear Catastrophe* (New York: St. Martin's, 1996), 115–38.

19. Adriana Petryna, "Sarcophagus: Chernobyl in Historical Light," *Cultural Anthropology* 10 (1995): 197.

20. Vitali Skliarov, *Chernobyl Was . . . Tomorrow: A Shocking Firsthand Account,* trans. Victor Batachov (Montreal: Les Presses d'Amerique, 1993), 8. Ten years later, even Gorbachev said he divides his life into "before and after" Chernobyl: Eileen O'Connor, "Chernobyl Survivors Recall a Decade of Death and Denial," *CNN Interactive World News,* 26 Apr. 1996.

21. The World Health Organization (WHO) estimates that the material released from the reactor was 200 times that of combined releases of the atomic bombs used on Hiroshima and Nagasaki. This figure has often been repeated. WHO, *Health Consequences of the Chernobyl Accident* (Geneva: WHO, 1995), 3.

22. In the context of the Chernobyl disaster, Newton Minow argues that new communication networks "make it difficult for governments to control data flow, unleashing an unprecedented volume of news and information across national borders." In Minow, "Introduction: Chernobyl and the New Age of Communications," in *Chernobyl: Law and Communication,* ed. Phillips Sands (Cambridge, U.K.: Grotius, 1988), xxi.

23. Abigail Trafford and Andrea Gabor, "Living Dangerously," *U.S. News & World Report,* 19 May 1986: 19.

24. Richard F. Mould, *Chernobyl: The Real Story* (New York: Pergamon, 1988), 180.

25. Frederik Pohl, *Chernobyl* (New York: Bantam, 1987), 119.

26. David R. Marples, *The Social Impact of the Chernobyl Disaster* (New York: St. Martin's, 1988), 4.

27. Richard Rhodes, *Nuclear Renewal: Common Sense About Energy* (New York: Viking, 1993), 92–99. In this section, Rhodes discusses Bernard Cohen's *The Nuclear Energy Option: An Alternative for the '90s* (New York: Plenum, 1990).

28. Thomas Hughes, *American Genesis: A Century of Invention and Technological Enthusiasm, 1870–1970* (New York: Viking, 1989), 470.

29. Alexander R. Sich, "Truth Was an Early Casualty," *Bulletin of the Atomic Scientists,* May–June 1996: 34.

30. Petryna, 196.

31. Marples, *Social Impact,* 167.

32. A. M. Petrosyants, former chairman of the Soviet State Committee on the Utilization of Atomic Energy, speaking at a 1986 press conference in Moscow. Qtd. in Sich, "Truth Was an Early Casualty," 32.

33. Read, 332.

34. A complete technical description of the scientific expedition and its findings can be found in Borovai and Sich, 1–32. Photographs of expedition workers in the reactor, and their rationales for working there, can be found in Francis X. Clines, "Into the White Hot Center," *New York Times Magazine,* 14 Apr. 1991: 35.

35. Eduard Pazukhin, qtd. in Robin Dixon, "Chernobyl, Mon Amour: Inside a Toxic Timebomb," *Sydney Morning Herald,* 13 Apr. 1996: A13.

36. Susan Benkelman, "Learning to Love a Disaster: Study of Chernobyl Enthralls Scientists," *Newsday,* 22 Nov. 1994: A25.

37. Stephen Baker and Carol Matlack, "Chernobyl: If You Can Make It Here," *Business Week,* 30 Mar. 1998: 168.

38. Sich, "Truth Was an Early Casualty," 41.

39. A full description of the team's findings can be found in OECD/NEA, *Sarcophagus Safety "94": The State of the Chernobyl Nuclear Power Plant Unit 4* (Paris: OECD, 1995).

40. A. F. Usatyj, "Generalized Results of Determination of Distribution of Major Gamma Radiation Sources in the Central Hall of the Sarcophagus, Recorded by Dosimetric Cords Using EPR Sensors," in *Sarcophagus Safety "94,"* 188.

41. Valentyn Kupny, qtd. in Rostislav Khotin, "Chernobyl 'Coffin' Sparks New Fear," *Herald* (Glasgow), 18 Sept. 1996: 12.

42. "Chernobyl Pledges Fall Short," BBC News, 21 Nov. 1997.

43. Elaine Monaghan, "Smiling Chernobyl Workers Greet Visitors," Reuters News Service, 18 June 1997; Vladimir Yemelyanenko, "Chernobylworld," *Moscow News,* 18 Apr. 1996: 1–2; Francis X. Clines, "As Villagers Return Unfazed, Chernobyl Aims for Tourists," *New York Times,* 4 Feb. 1991: A1, A3; "Next Stop: Chernobyl 'Dead Zone,'" *Chicago Tribune,* 28 Apr. 1994: sports final ed., 24.

44. The number by 1990 had reached a half-million according to E. Edward Miller and Ruby M. Miller, *Environmental Hazards: Radioactive Materials and Wastes* (Santa Barbara, Calif.: ABC-CLIO, 1990), 55.

45. Dixon, A13.

46. Mary Myco, "A Legacy of Wormwood," *Newsday,* 7 Apr. 1996: A.

47. Dave Carpenter, "Deadly Cloud of Uncertainty Still Hangs over Chernobyl," AP Wire Service, 14 Apr. 1996.

48. David Lascelles, "Journey to Centre of World's Worst Nuclear Disaster," *Financial Times* (London), 30 Oct. 1995: 2; Richard D. North, "Living with Catastrophe," *Independent* (London), 10 Dec. 1995: 4.

49. Vladimir Voina, "Perspective on Chernobyl: Revisiting the Unthinkable," *Los Angeles Times,* 8 May 1995: B5.

50. Marples, *Social Impact,* 48.

51. World Health Organization, 15.

52. EC/IAEA/WHO, "One Decade after Chernobyl: Summing up the Consequences of the Accident," Proceedings of an International Conference on One Decade after Chernobyl, 8–12 Apr. 1996, Vienna (Vienna: The Agency, 1996), 10.

53. Lyubov Sirota, "Radiophobia," in *Life on the Line: Selections on Words and Healing,* ed. Sue Brannen Walker and Rosaly Demaios Roffman (Mobile, Ala.: Negative Capability Press, 1992), 619.

54. Dawson, 29.

55. Ibid., 82. A little under half of the Ukraine's energy comes from nuclear power as of this writing.

56. A ten-year assessment by the OECD National Energy Agency's Committee on Radiation Protection and Public Health states that "the Bryansk-Belarus spot, centered 200 km to the North-northeast of the reactor, was formed on 28–29 April as a result of rainfall on the interface of the Bryansk region of Russia and the Gomel and Mogilev regions of Belarus. The ground depositions of caesium-137 in the most highly contaminated areas in this spot were comparable to the levels in the Central spot and reached 5,000 kBq/m^2 in some villages." *Chernobyl, Ten Years On: Radiological and Health Impact* (OECD/NEA, 1995), Ch. 2, available online at http://www.nea.fr/html/rp/chernobyl/ (18 Sept. 1998). Medvedev states that the evacuation zones were "arbitrary" (144).

57. Voina, B5.

58. "'There Is a Silent Enemy Lurking': In the Aftermath of a Nightmare, an Eerie Calm Prevails," *Time,* 23 June 1986: 49.

59. Steven Strasser, "In Chernobyl's Grim Shadow," *Newsweek,* 29 June 1987: 38.

60. Patty Reinert, "Stark Images of the Chernobyl Disaster," AP Wire Service, 26 June 1990. See also Judy Keen, "Chernobyl: Nuclear Reactor Tragedy Left Scars on Land, Lives and 'Will Never Be Over,'" *USA Today,* 17 Sept. 1991: 6A; and Lisa Trei, "Visit to a Dying Town Near Chernobyl Plant," *San Francisco Chronicle,* 24 Sept. 1991: A1.

61. Nova, *Back to Chernobyl* (Stamford, Conn.: Vestron Video, 1989).

62. O'Connor.

63. Carpenter.

64. Glenn Alan Cheney, *Journey to Chernobyl: Encounters in a Radioactive Zone* (Chicago: Academy Chicago, 1995), 154.

65. Laura Spinney, "Return to Paradise," *New Scientist,* 20 July 1996: 28.

66. Donald Bruce, "Living Death of a Ghost Town," *New Scientist,* 18 May 1996: 53.

67. Robert Bazell, "Red Forest," *New Republic,* 30 May 1988: 11.

68. Sally Lilienthal, "Thoughts in a Soviet Wasteland," *San Francisco Chronicle,* 27 Apr. 1989: A29. Lilienthal is the founder of Ploughshares, an antinuclear organization.

69. Michael Willard, "In Chernobyl, the Geiger Counter Continues to Buzz," *Charleston Daily Mail,* 22 Mar. 1995: A5.

70. Robert Peter Gale and Thomas Hauser, *Final Warning: The Legacy of Chernobyl* (New York: Warner, 1988), 50.

71. Gale and Hauser, 62.

72. Gale's visits were funded by Armand Hammer, who had long-standing business ties with the Soviets. Hammer's role is also presented as heroic, if somewhat eccentric, in *Final Warning.* In the film, he is portrayed by Jason Robards as loud and abrasive.

73. Robert Gale, Interview, *Omni,* Oct. 1987: 114.

74. Thirteen bone marrow transplants were performed in all, with two of those patients surviving. A description of the controversy over the transplants, and Gale's role in it, can be found in Marples, *Social Impact,* 34–35.

75. "Grim Lessons at Hospital No. 6," *Time,* 26 May 1986. See also Gale's description of his work in "Medical Response to Nuclear Accidents," in Robert Peter Gale and Richard E. Champlin, eds., *Bone Marrow Transplantation: Current Controversies,* Proc. of a Sandoz-UCLA Symposium, 6–12 Mar. 1986 (New York: Alan R. Liss, 1989), 671–78.

76. Gale, Interview, 120.

77. Robert Peter Gale, "USSR: Follow-Up after Chernobyl," *Lancet* 335 (1990): 401.

78. Philosopher Arthur O. Lovejoy identified what he called the "principle of plenitude" in a Western tradition stemming from Plato. It is made up of two notions: that the universe has a complete diversity of living things stemming from a "perfect and inexhaustible Source" and that "no genuine potentiality of being can remain unfulfilled." *The Great Chain of Being: A Study of a History of an Idea* (Cambridge, Mass.: Harvard University Press, 1964), 52. I have often found this tradition operating in disaster narratives that posit the disaster as both a locus of total destruction and a source of endless diversity, including the diversity of narrative itself.

79. Peter Gould, *Fire in the Rain: The Democratic Consequences of Chernobyl* (Cambridge, U.K.: Polity, 1990), xi.

80. V. K. Savchenko, *The Ecology of the Chernobyl Catastrophe: Scientific Outlines of an International Programme of Collaborative Research,* Man and the Biosphere Series 16 (Paris: UNESCO, 1995), xvii.

81. Ibid., 2.

82. The health status of living organisms in the zone is still very uncertain, and research

is by no means complete. Biological studies have been hampered by political difficulties in accessing the site and by state secrecy and bumbling in record keeping. Furthermore, radiation effects take decades to emerge in a single organism and may have consequences that stretch over many generations. In other words, no one yet knows the full effects of this disaster, and we may not know for a very long time. My description of the ecological conditions in the zone is based on the work of UNESCO's Chernobyl Ecological Science Network, summarized by geneticist V. K. Savchenko in *The Ecology of the Chernobyl Catastrophe,* and on a study of small mammals in the region conducted by a team from Texas Tech. See Robert J. Baker et al., "Small Mammals from the Most Radioactive Sites Near the Chornobyl Nuclear Power Plant," *Journal of Mammalogy* 77 (1996): 155–70.

83. The area around the DOE's Savannah River Plant in South Carolina, home to five nuclear reactors, has also been converted into an "ecological zone." Jeffrey Cohn describes the "awe and excitement" of ecologists at the site; its " stately long-leaf and loblolly pines, sweet gums, oaks, and bald cypress remind scientists and visitors alike of how the southeastern coastal plain may have looked centuries ago." In "A National Natural Laboratory," *Bioscience* 44 (1994): 727.

84. Cornelia Hesse-Honegger, "Painting Mutations," *Geographical,* Nov. 1992: 15; Hesse-Honegger, *The Future's Mirror* (Newcastle, U.K.: Locus+, 1996).

85. Hesse-Honegger, "Painting Mutations," 18.

86. Holland Cotter, Review of Cornelia Hesse-Honegger at the Swiss Institute, *New York Times,* 24 Sept. 1993: 30; William Thompson, "The Line between Art and Science," *Geographical,* Nov. 1992: 18–19.

87. Locus+, press release, 24 Apr. 1997.

88. Charles Leroux, "Chernobyl's Radiant Beauty: Their Greatest Threat Gone, Plants and Wildlife Flourish at the Site of the World's Worst Nuclear Disaster," *Chicago Tribune,* 20 June 1997, sec. 5, p. 1. This article on the work of Baker and the Texas Tech scientific team of which he was part was widely syndicated. See also Mary Mycio, "Minus Humans, Wildlife Thrives in Chernobyl Area," *Los Angeles Times,* 26 Feb. 1996: A1; and Scott Shane, "Chernobyl's Persistent Fallout," *Baltimore Sun,* 14 Apr. 1996: 2A.

89. Karen F. Schmidt, "The Truly Wild Life around Chernobyl: Many Animals Are in Evolutionary Overdrive," *U.S. News & World Report,* 17 July 1995: 51.

90. Alan Weisman, "Journey through a Doomed Land: Exploring Chernobyl's Still-Deadly Ruins," *Harper's,* Aug. 1994: 45. The birds are significant because, in the year after the Chernobyl disaster, few birds could be found in the area.

91. David Strong, *Crazy Mountains: Learning from Wilderness to Weigh Technology* (Albany: SUNY Press, 1995), 54.

92. John D. Cramer, "Chernobyl Leaves a Lonely, Quiet Land: Neighbors Cope with Solitude, Reminders of 1986 Accident's Radiation," *Milwaukee Journal Sentinel,* 16 Feb. 1997: 1.

93. Jim Cheney, "Postmodern Environmental Ethics: Ethics as Bioregional Narrative," *Environmental Ethics* 11 (1989): 117.

94. Benedict Anderson explains that colonization and the rise of nationalism depended on these three interlinking activities: the census, the map, and the museum. I am borrowing his categories to describe the resistance of the Chernobyl zone to colonization. See *Imagined Communities: Reflections on the Origin and Spread of Nationalism* (New York: Verso, 1991), 184.

95. Marples, *Belarus,* 39.

96. Bruno Latour explains that new sociotechnical aggregations, like scientific laboratories, socialize nonhuman entities by transforming them into codes, books, and programs of action. In his example of yeast, he describes how earlier social aggregations had domesticated its material properties for brewing. But more recently, yeast has been used by scientists mapping its genome so that it behaves and communicates as a code, "no longer retaining any of its material quality, its outsiderness. It has been swallowed within the collective" (798). In "Pragmatogonies: A Mythical Account of How Humans and Nonhumans Swap Properties," *American Behavioral Scientist* 37 (1994): 791–808.

97. Information about health monitoring comes from the World Health Organization's summary report of its activities in the zone, *Health Consequences of the Chernobyl Accident*, 5–13.

98. David R. Marples, "The Decade of Despair," *Bulletin of the Atomic Scientists*, May–June 1996: 24.

99. In 1996, Andrey M. Serdyuk of the Ukrainian Ministry of Health stated that 737 cases of thyroid cancer had been found among children since the Chernobyl accident. In Shunichi Yamashita and Yoshisada Shibata, *Chernobyl: A Decade*, Proc. of the Fifth Chernobyl Sasakawa Medical Cooperative Symposium, 14–15 Oct. 1996, Kiev (New York: Elsevier, 1997), ix. The Chernobyl Sasakawa Medical Cooperative brought Japanese scientists who had been working with Hiroshima and Nagasaki survivors to the Ukraine, where they helped conduct a collaborative health screening of Chernobyl children. Although they focused mostly on thyroid cancers, some participants in the symposium speculated on the overall breakdown of children's health, suggesting that it was the result of radiation exposure.

100. Yuri E. Dubrova et al., "Human Minisatellite Mutation Rate after the Chernobyl Accident," *Nature* 380 (1996): 683–87.

101. Cramer, 1.

102. Dave Carpenter, "Chernobyl's Ghost Haunts the Ukraine," *Los Angeles Times*, 14 Apr. 1996: A1.

103. Andrew Melnykovych, "Old People Return to Villages Near Chernobyl," Gannett News Service, 21 Apr. 1996: S12.

104. Glenn Alan Cheney, 165.

105. Masha Gessen, "After Technology," *Wired*, Mar. 1996, 145.

106. Julian Borger, "Back Inside the Dead Zone," *World Press Review*, Aug. 1994: 11.

107. Valya Kovtunenko, qtd. in Melnykovych. For other accounts of the elderly homesteaders, see Charles Strouse, "Living in the Shadow of Chernobyl," *Fort Lauderdale Sun-Sentinel*, 30 Nov. 1996: 1A; "Some Chased by Chernobyl Return to Die," *Des Moines Register*, 21 Apr. 1996: 9.

108. Andrew Nagorski, "'The Zone of Alienation,'" *Newsweek*, 22 Apr. 1996: 54.

109. Felicity Barringer, "Chernobyl: Five Years Later, the Danger Persists," *New York Times Magazine*, 14 Apr. 1996: 28.

110. Harvey Wasserman writes, "The dead zone has become a post-apocalyptic no man's land for drifters and lost souls who find a haven of sorts in abandoned houses and empty gardens," in "In the Dead Zone: Aftermath of the Apocalypse," *Nation*, 29 Apr. 1996: 18.

111. Anna Krasyukova, qtd. in Larisa Sayenko, "War Refugees Find Paradise in Chernobyl Zone," Reuters News Service, 4 Sept. 1995. See also James Rupert, "Life is Pretty Bad," *Washington Post*, 18 Apr. 1995: A19; and Larisa Sayenko, "Chernobyl 10 Years Later: Tragedy's Toll Mounts," *The Current Digest of the Post-Soviet Press*, 22 May 1996: 14.

112. Tim Hundley, "Forbidden Landscape," *Chicago Tribune*, 21 Apr. 1996: 1, 9.

113. Paul Salopek, "Latest Hot Zone for Diseases May Be Right Out Your Window," *Chicago Tribune,* 21 Apr. 1996: 1, 8.

114. James Rupert, "Nuclear Blight Invades Minds as Well as Bodies," *Washington Post,* 18 Apr. 1996: A18.

115. Reinert.

116. "Children of the Gulag," *Mail on Sunday,* 18 Feb. 1996: 51.

117. Warren Christopher, Address, "American Diplomacy and the Global Environmental Challenges of the 21st Century." Stanford University, Palo Alto, Calif., 9 Apr. 1996.

118. Saving Chernobyl children is a highly visible activity in Ireland, a nuclear-free country. Rock star Bono and his wife, Ali Hewson, have participated in Adi Roche's controversial Chernobyl Children's Project, which has brought more than a thousand children to Ireland to strengthen their immune systems. Ireland's claims to providing a clean environment for these children are certainly more compelling than those of the nuclear states.

119. Nonna Novik, qtd. in Marylou Tousignant, "Only Glow Is Love for Families Hosting Chernobyl's Children," *Los Angeles Times,* 13 July 1997: A28. See also Louise Mengelkoch, "Learning to Live after Chernobyl," *Seattle Times,* 24 July 1995: B1.

120. Tousignant, A28.

121. Pam Starr, "Russian Children Find Beach 'Really Cool,'" *Virginia Pilot,* 23 July 1997: B4; Elizabeth Carvlin, "Belarussian Children in Town for Good Food, Health Care, TLC," *Chapel Hill Herald,* 1 July 1997: 1; Mia Taylor, "Leaving American Life, Family; Child from Chernobyl Prepares to Go Home," *Patriot Ledger,* 14 Feb. 1998: 1.

122. Elaine Laporte, "Chernobyl Survivors Celebrate First Seder American-Style," *Jewish Bulletin of Northern California,* 1995, available online at http://www.jewish.com/bk960412/usceleb.htm (14 Nov. 1997).

123. "U.S. Sperm Bank to Aid Ukraine," *Phoenix Gazette,* 27 Jan. 1996: A2.

124. Joel Jacobson, "Chernobyl Fallout Bears a Blossom," *Halifax Herald,* 28 June 1996: 5.

125. Mikhail Sazonenko, qtd. in Hundley, 9.

126. Vasiliy Osipovich Kotestky, in *Testimonies: Chernobyl Papers No. 1* (Amsterdam: Greenpeace International, 1996), 10.

127. Anton Antonovich Vulchin, in *Testimonies,* 23.

128. Alla Yaroshinskaya, *Chernobyl: The Forbidden Truth,* trans. Michèle Kahn and Julia Sallabank (Oxford, U.K.: Jon Carpenter, 1994), 27.

129. Ibid., 28.

Chapter 3: Radioactive Body Politics

1. Rod Serling, "Time Enough to Last," *Twilight Zone,* dir. John Brahm, 20 Nov. 1959.

2. Andrew Holleran, "Ground Zero," in *Ground Zero* (New York: Morrow, 1988), 22.

3. Paula Treichler, "AIDS, Homophobia, and Biomedical Discourse; An Epidemic of Signification," *October* 43 (1987): 31.

4. Holleran, "Reading and Writing," in *Ground Zero,* 12.

5. Ibid.

6. Ibid., 13.

7. Holleran, "Ground Zero," 20, 28.

8. Ibid., 26.

9. Qtd. In Robert J. Lifton, *History and Human Survival* (New York: Random House, 1970), 161.

10. Michael S. Sherry mentions that Chris Glaser has written an essay titled "AIDS and the A-Bomb Disease," which compares the gay community's exposure to HIV with the *hibakusha's* experience. See Sherry, "The Language of War in AIDS Discourse," in *Writing AIDS: Gay Literature, Language, and Analysis,* ed. Timothy F. Murphy and Suzanne Poirier (New York: Columbia University Press, 1993), 43.

11. Jonathan Schell, *The Fate of the Earth* (New York: Knopf, 1982), 120; Helen Caldicott, *Missile Envy: The Arms Race and Nuclear War* (New York: Bantam, 1986), 312.

12. The entire issue of *Diacritics* 14 (Summer 1984) was devoted to nuclear criticism. It was followed by a special issue of *Papers in Language and Literature* 26 (Winter 1990) with an introduction by William J. Scheick, "Nuclear Criticism: An Introduction," which discussed nuclear criticism's endeavor to locate a "renewed or new ethical sensibility" (5). Peter Schwenger's *Letter Bomb: Nuclear Holocaust and the Exploding Word* (Baltimore, Md.: Johns Hopkins University Press, 1992) hopes for a "new homeland" constructed through unstable and intertextual writings about nuclear war (xiv).

13. Larry Kramer, *Report from the Holocaust: The Making of an AIDS Activist* (New York: St. Martin's, 1989), 189.

14. Larry Kramer, Interview, *Playboy* 40 (Sept. 1993): 61.

15. Emmanuel Dreuilhe, *Mortal Embrace,* trans. Linda Coverdale (New York: Hill & Wang, 1988), 19.

16. Charles Perrow and Mauro F. Guillén, *The AIDS Disaster: The Failure of Organizations in New York and the Nation* (New Haven, Conn.: Yale University Press, 1990), 181–83.

17. Sherry, 40, 49–50.

18. Stephen Fried, "Cocktail Hour," *Washington Post,* 18 May 1997: W10; Ron Cohen, "The Changing of the Guard," *Science* 28 (1996): 1876.

19. Richard Dellamora, *Apocalyptic Overtures: Sexual Politics and the Sense of an Ending* (New Brunswick, N.J.: Rutgers University Press, 1994), 28.

20. Scheick, 6.

21. Holleran, "Oceans," in *Ground Zero,* 226.

22. Randy Shilts, *And the Band Played On: Politics, People, and the AIDS Epidemic* (New York: St. Martin's, 1987).

23. Ibid., 260.

24. Schwenger, 25.

25. Ibid., 30–38.

26. Shilts, 157.

27. Ibid., 12.

28. Ibid., 21.

29. Ibid., 12.

30. Ibid., 22.

31. Ibid., 147.

32. Ibid., 12.

33. Ibid., 568.

34. Gillian Brown, "Nuclear Domesticity: Sequence and Survival," in *Arms and the Woman: War, Gender, and Literary Representation,* ed. Helen M. Cooper, Adrienne Auslander Munich, and Susan Merrill Squier (Chapel Hill: University of North Carolina Press, 1989), 283–302.

35. Shilts, 605.

36. Holleran, "Reading and Writing," 18.

37. The first installment of Tony Kushner's play, *Angels in America: Millennium Approaches*, was first performed by the Los Angeles Center Theater Group/Mark Taper Forum in May 1990 and later played at London's National Theater and on Broadway. It has been published by the Theatre Communications Group (New York, 1993). Part II, *Angels in America: Perestroika*, opened on Broadway in Nov. 1993.

38. *Angels in America: Millennium Approaches*, 118.

39. Richard Preston's bestselling *Hot Zone* (New York: Anchor, 1994) articulated broadening fears of Level 4 viruses such as ebola, first sounded in the 1980s with the recognition of HIV.

40. Robert Freitas, "The Birth of the Cyborg," in *Robotics*, ed. Marvin Minsky (Garden City, N.Y.: Anchor/Doubleday, 1985), 150.

41. David M. Rorvik, *As Man Becomes Machine: The Evolution of the Cyborg* (Garden City, N.Y.: Doubleday, 1971); D. S. Halacy, *Cyborg: Evolution of the Superman* (New York: Harper & Row, 1965).

42. Jerry Adler, "Reading God's Mind," *Newsweek*, 13 June 1988: 56.

43. Michael White and John Gribbin, *Stephen Hawking: A Life in Science* (New York: Plume, 1993), 133.

44. Shawn Rosenheim, "Extraterrestrial: Science Fictions in 'A Brief History of Time' and 'The Incredible Shrinking Man,'" *Film Quarterly* 22 (June 1995): 15.

45. Allucquere Rosanne Stone, *The War of Desire and Technology at the Close of the Mechanical Age* (Cambridge, Mass.: MIT, 1995), 5.

46. White and Gribbin, 139.

47. Cynthia Fuchs suggests that the cyborg's performance represents a crisis of gendered (especially masculinist) identity in "'Death Is Irrelevant': Cyborgs, Reproduction, and the Future of Male Hysteria," *Genders* 18 (Winter 1993): 115.

48. David Wojnarowicz, *Close to the Knives: A Memoir of Disintegration* (New York: Vintage, 1991), 117.

49. William S. Burroughs, *The Wild Boys*, in *Three Novels* (New York: Grove Weidenfeld, 1980), 504.

50. John Carlin, "David Wojnarowicz: As the World Turns," in *David Wojnarowicz: Tongues of Flame*, ed. Barry Blinderman (Normal: University Galleries of Illinois State University, 1990), 29.

51. Wojnarowicz, 37.

52. Ibid., 38.

53. Ibid., 63.

54. Ibid.

55. Andrew Ross, *Strange Weather: Culture, Science, and Technology in the Age of Limits* (New York: Verso, 1991), 100.

56. Wojnarowicz, 24.

57. Felix Guattari, "David Wojnarowicz," *Globe E-Journal*, issue 6, available online at http://www.arts.monash.edu.au/visarts/globe/issue6/davwoj.html (25 Sept. 1998).

58. E. M. Forster, "The Machine Stops," in *The Collected Tales of E. M. Forster* (New York: Knopf, 1964), 144–98; Judith Merril, "That Only a Mother," *Women of Wonder: Science Fiction Stories by Women about Women*, ed. Pamela Sargent (New York: Vintage, 1975), 5–17; Daniel F. Galouye, *Dark Universe* (Boston: Gregg, 1976); Harlan Ellison, "I Have

No Mouth, and I Must Scream," *Alone Against Tomorrow* (New York: Macmillan, 1971), 15–32.

59. Paula Treichler and Lisa Cartwright, Introduction, *Camera Obscura* 28 (Jan. 1992), 15–32. See also Alexandra Juhaxz on the Women's AIDS Video Action Project, "WAVE in the Media Environment: Camcorder, Activism and the Making of HIV TV," *Camera Obscura* 28 (Jan. 1992): 135–50.

Chapter 4: Oil and Water

1. Peter Marks, "Hazelwood's Year: 'Joe Schmoe' to Villain," *Anchorage Daily News*, 25 Mar. 1990: B1; Joe Hunt, "It's a Terrible Tragedy: It'll Affect Me Like It Will Affect Everybody Else for a Long Time," *Anchorage Times*, 25 Mar. 1990: A1.

2. *Waterworld*, dir. Kevin Reynolds, with Kevin Costner, Dennis Hopper, and Jeanne Tripplehorn, Universal Studios, 1995.

3. *MacNeil/Lehrer Newshour*, PBS, 30 Mar. 1989; Interview with Lee Raymond, David Brinkley, *This Week with David Brinkley*, ABC, 2 Apr. 1989. Statement by Frank Iarossi, Exxon Press Release on Hazelwood's firing, Houston, Texas, 30 Mar. 1989.

4. Coast Guard Commandant Paul Yost made this statement at a 30 March 1989 press conference in Washington, D.C., after a fact-finding mission to Prince William Sound.

5. Bill Bleyer, "On the Water: Maritime Lawyer's Travels, Travails," *Newsday*, 27 Feb. 1994: 27.

6. *State of Alaska vs. Joseph J. Hazelwood*, 3ANS89–7217, 3ANS89–7218 (Alaska Superior Court 1989).

7. Michael Lewis, "Exxon Kills Prince William: A Cost of the Infernal Combustion Engine," *Earth First!* 1 May 1989: 7.

8. Art Davidson, *In the Wake of the* Exxon Valdez: *The Devastating Impact of the Alaska Oil Spill* (San Francisco: Sierra Club, 1990), xi.

9. Paul Hawken, "The Ecology of Commerce," *Inc.*, Apr. 1992: 93.

10. John Keeble, *Out of the Channel: The* Exxon Valdez *Oil Spill in Prince William Sound* (New York: HarperCollins, 1991), 268.

11. Qtd. in Davidson, 109.

12. Joe Hunt, "Focus of Trial May Miss Mark," *Anchorage Times*, 28 Jan. 1990: A1.

13. Hazelwood was convicted in 1990 of a misdemeanor, negligently discharging oil, and sentenced to 1000 hours cleaning oiled beaches and $50,000 restitution to the state. This conviction has been overturned twice in Alaska courts. As of this writing prosecutors are still pursuing charges against Hazelwood.

14. Hunt, "Terrible Tragedy," A1, A8.

15. Testimony of Joseph Hazelwood, *State vs. Joseph J. Hazelwood*, A89–0095 CIV (HRM) (U.S. District Court for the District of Alaska 1994). Qtd. In "Natives Fight Removal from 'Valdez' Oil Spill Case," *All Things Considered*, NPR, 15 May 1994. See also Natalie Phillips, "Hazelwood Says Shock of Spill Made Him Sick," *Anchorage Daily News*, 13 May 1994: A1, A8.

16. Hazelwood's smoking is mentioned in "Judge Refuses to Drop Charges against Skipper." *Anchorage Times*, 29 Jan. 1990: A1; and Marks, B1.

17. Pope Brock, "Lost at Sea—And on Land," *Life*, 1 Feb. 1990: 78.

18. Ibid., 83. Don Hunter explains that *Life* "accepted Hazelwood's version of key events in the grounding without offering contradictory evidence and without talking to anyone

other than Hazelwood." Furthermore, he says that restrictions imposed on interviewers by Hazelwood's lawyers would not allow them to discuss the accident or his drinking. The strict protection of information because of legal considerations certainly stood in the way of any emergence of truth concerning the disaster, not only in this instance but throughout the discourse on the *Exxon Valdez* oil spill. Hunter, "Lawyers Keep Tight Media Lid," *Anchorage Daily News,* 3 Feb. 1990: D1, D3.

19. Brock, 83.

20. Qtd. in William C. Rempel, "Hazelwood Tells of Images That Still Haunt Him," *Los Angeles Times,* 25 Mar. 1990: A1.

21. Hunt, "Terrible Tragedy," A8.

22. Ibid.

23. Eric Nalder, *Tankers Full of Trouble: The Perilous Journey of Alaskan Crude* (New York: Grove, 1994), 80–83. Some, including Nalder, partly blame worker fatigue for the spill, arguing that cutbacks to tanker crews made for disaster. Third mate Gregory Cousins was reported to have been very tired during his shift, provoking Walter Parker, chairman of the Alaska Oil Spill Commission, to say, "The shipping industry has come late to the automation game and would do well to examine some of the dark sides of automation with reduced crews." In John Foley, "More Spills Predicted Unless Changes Made," *Anchorage Times,* 8 Apr. 1990: C2.

24. *State of Alaska vs. Joseph J. Hazelwood* (1989).

25. Ken Gross, "Up Front: He Devastated Alaska, but Capt. Jeff Hazelwood, Friends Say, Is an Environmentalist," *People,* 24 Apr. 1989: 48.

26. Richard Behar, "Joe's Bad Trip," *Time,* 24 July 1989: 44.

27. Ibid., 45.

28. Natalie Phillips, "Hazelwood Details His Struggle with Booze: *Exxon Valdez* Captain Says Many Company Officials Knew about His Drinking," *Anchorage Daily News,* 11 May 1994: A1.

29. Joe Hunt, "No Setback to Their Cause, Groups Claim," *Anchorage Times,* 23 Mar. 1990: A1. Evans often echoed the work of Charles Perrow. Later she would say, "Exxon and Hazelwood are just two agents in a very complicated and very flawed system of extraction and transportation of petroleum." Qtd. in Paul A. Witteman, "First Mess Up, Then Mop Up: Hazelwood Is Ordered to Help Cleanse Alaska's Shoreline," *Time,* 2 Apr. 1990: 22.

30. O. B. Hardison, Jr., *Disappearing through the Skylight: Culture and Technology in the Twentieth Century* (New York: Penguin, 1989), 192–93.

31. Laurie Stone, "Factor X and the Big Spill," *Tikkun,* Mar.–Apr. 1990: 72.

32. Keeble, 16.

33. Joni Seager, *Earth Follies: Coming to Feminist Terms with the Global Environmental Crisis* (New York: Routledge, 1993), 101–2. Some have argued more broadly that the big technological disasters of the 1980s, such as Bhopal, Chernobyl, and the *Exxon Valdez,* sparked a renewed interest in environmental movements, including ecofeminism. Dorothy Kleffel, "Environmental Paradigms: Moving toward an Ecocentric Perspective," *Advances in Nursing Science* 18 (June 1996): 1–10; Paul Shrivastava, "Castrated Environment: Greening Organizational Studies," *Organization Studies* 15 (1994): 705–26.

34. Charles Wohlforth, "DEC Chief Dennis Kelso Enjoys a Champion's Status in Cordova," *Anchorage Daily News,* 30 Apr. 1989: A1.

35. Lisanne Renner, "Pristine Past Just an Echo in Prince William Sound," *Orlando Sentinel Tribune,* 25 Mar. 1990: A1.

36. The Oil Spill Information Center contains unedited footage of interviews with cleanup workers who describe their jobs and dissatisfactions.

37. Ernest Piper, *The* Exxon Valdez *Oil Spill: Final Report, State of Alaska Response* (Anchorage: Alaska Department of Environmental Conservation, 1993), 95. See also Brian O'Donoghue, *Black Tides: The Alaska Oil Spill* (Anchorage: Alaska Natural History Association, c. 1989), 28–29.

38. Bob Ortega, "Day-by-Day Account of the Spill Shows Evolving Tragedy," *Homer News,* 29 Dec. 1989: 13.

39. Red Kvarford, qtd. in *The Day the Water Died: A Compilation of the November 1989 Citizens Commission Hearings on the* Exxon Valdez *Oil Spill,* ed. Thea Levkovitz (Anchorage: National Wildlife Federation, 1990), 31.

40. Sharon McClintock, "Alaska Oil Spill Commission: Oiled Communities Response Investigation Report," in Alaska Oil Spill Commission, *Spill: The Wreck of the* Exxon Valdez (State of Alaska, 1990), 27.

41. Lawrence Rawl, press conference, 18 Apr. 1989, in *Legi-Slate Report,* 24 Apr. 1989: 38.

42. Bridget Milligan, qtd. in *The Day the Water Died,* 28.

43. Kathryn Kinnear, qtd. in *The Day the Water Died,* 28.

44. Piper, 61.

45. Rawl, press conference, 37.

46. "Exxon Cited for Orwellian Tactics," *Public Relations Journal,* Feb. 1990: 11.

47. Mike Berg, qtd. in *The Day the Water Died,* 31.

48. *Prince William Sound, Alaska: One Year After,* brochure, Exxon, c. 1990.

49. Joyce Thompson, "Alki Beach," in *Season of Dead Water,* ed. Helen Frost (Dallas: Breitenbush, 1990), 34.

50. Ann F. Lewis, "Alaskan Housekeeping," *Ms.,* June 1989: 76.

51. Tom Horton, "Paradise Lost," *Rolling Stone,* 14 Dec. 1989: 173.

52. Alaska Visitor's Association, "Legendary Beauty," Anchorage, Bradley Advertising, 1989.

53. Lynn Chrystal, "Presentation to the Resource Development Council of Alaska," *Valdez Vanguard,* 20 Dec. 1989: A7.

54. Local reports in the months before the mayor's speech should have alerted him to the problem. See Sharon Ann Jaeger, "Counselors Say Stress Levels in Community Are Rising," *Valdez Vanguard,* 5 July 1989: A-1, A-16; Terry Wilson, "Psychologist Says Stress Is Byproduct of Oil Spill Cleanup," *Valdez Vanguard,* 25 Oct. 1989: A6. Other locally funded studies that indicate a dramatic rise in domestic violence include Valdez Counseling Center, unpublished paper, "The Stress Impact of the Valdez Oil Spill on the Residents of Cordova and Valdez," Valdez, Alaska, June 1990. Sharon K. Araji, unpublished paper, "The *Exxon-Valdez* Oil Spill: Social, Economic, and Psychological Impacts on Homer," presented to the Homer Community, 22 Apr. 1992; Sharon K. Araji, unpublished paper, "The *Exxon Valdez* Spill: Social Economic, and Psychological Impacts on Seldovia," presented to Seldovia Community, 27 Apr. 1992; and Lawrence Palinkas et al., "Social, Cultural, and Psychological Impacts of the *Exxon Valdez* Oil Spill," *Human Organization* 52 (Spring 1993): 1.

55. Ray Voley, "Bird Volunteers Should Have a 'Knack,'" *Kodiak Daily Mirror,* 10 Apr. 1989.

56. Laura Van Tuyl, "Otter Volunteer No Longer Feels Helpless," *Los Angeles Times,* 23 Sept. 1989, Part 5; p. 19.

57. Elizabeth Pulliam, "The Otter Ward," *Anchorage Daily News,* 11 Apr. 1989: H1.

58. Suzanne Marinelli, qtd. in Davidson, 163. This anthropomorphizing became a cliché among rescue workers. See S. F. Loshbaugh, "Little Jakolof Doing It the Otter Way," *Homer News,* 15 June 1989; Voley.

59. Horton, 159.

60. Jon R. Luoma, "Terror and Triage at the Laundry," *Audubon,* Sept. 1989: 96.

61. Tektite's Mission 6 team also included biologists Renate Schlenz True and Alina Szmant, ecologist Ann Hartline, and habitat engineer Ann Lucas. The press described them as the "aquanettes." The navy, the Department of the Interior, NASA, and General Electric funded the project, mostly for research on people confined to small isolated quarters.

62. Earle, qtd. in Peggy Orenstein, "Champion of the Deep," *New York Times Magazine,* 23 June 1991: 16.

63. Sylvia Alice Earle, *Sea Change: A Message of the Oceans* (New York: Putnam, 1995), 266.

64. Ibid., 268–69.

65. Wallace White, "Her Deepness," *New Yorker,* 2 July 1989: 42.

66. Riki Ott, *Sound Truth: Exxon's Manipulation of Science & the Significance of the Exxon Valdez Oil Spill* (Anchorage: Greenpeace, Alaska, 1994), 8.

67. Bill Sherwonit, "People Who Make a Difference: Riki Ott," *National Wildlife* 31 (Oct.–Nov. 1993): 47.

68. Stone, 72.

69. Wildlife biologists Chuck Monnett and Lisa Rotterman, qtd. in Davidson 158.

70. Horton, 170.

71. Keeble, 263.

72. Luoma, 97.

73. Ibid., 100.

74. Ken Hill, qtd. in Larry B. Stammer, "Alaskan Spring a Season of Death: Outlook for Recovery from Oil Spill Unclear," *Los Angeles Times,* 23 Apr. 1989: 1.

75. Keeble, 169.

76. Ynestra King, "The Ecology of Feminism and the Feminism of Ecology," in *Healing the Wounds: The Promise of Ecofeminism,* ed. Judith Plant (Toronto: Between the Lines, 1989), 22–23.

77. Ruth Schwartz Cowan, *More Work for Mother: The Ironies of Household Technology from Open Hearth to the Microwave* (New York: Basic Books, 1983).

78. Randall Davis, qtd. in John Wilkins, "Otters Find Ray of Hope amid Oil Slick; Sea World Expert in Alaska to Save Stricken Animals," *San Diego Union Tribune,* 7 May 1989: A3.

79. Carolyn McCollum, qtd. in Charles McCoy, "Heartbreaking Fight Unfolds in Hospital for Valdez Otters," *Wall Street Journal,* 20 Apr. 1989: 1.

80. Estimates of otter deaths now range in the thousands. 357 were rescued; 197 survived. Of those saved, each animal cost Exxon $80,000. Many sources agree that otters have recovered well, unlike some other wildlife populations. For biological assessments of postspill conditions for otters, see T. R. Loughlin, B. E. Ballachey, and B. A. Wright, "Overview of Studies to Determine Injury Caused by the *Exxon Valdez* Oil Spill to Marine Mammals," *Proceedings of the* Exxon Valdez *Oil Spill Symposium,* Vol. 18 (Bethesda, Md.: American Fish Society, 1996), 798–808; R. A. Garrott, L. L. Eberhardt, and D. M. Burn, "Mortality of Sea Otters in Prince William Sound Following the *Exxon Valdez* Oil Spill," *Marine Mammal Science* 9 (1993): 343–59.

81. "Effort to Save Otters in Spill Is in Dispute," *New York Times,* 21 Apr. 1990: 8.

82. Rick Steiner, "Probing an Oil-Stained Legacy," *National Wildlife* 31 (Apr.–May 1993): 6. Cordova mayor Donald Moore also wrote a national commentary referring to Alaska as a ruined Shangri-La in "Let the World Bear Witness to Alaska's Woe," *Newsday,* 26 Apr. 1989: 73. In a piece on the impact of the spill on the Aleut village of English Bay, the *New York Times* referred to it as a "sort of Alaskan version of Shangri-La." Timothy Egan, "Oil Cleanup Brings Sad Prosperity to Village," *New York Times,* 18 Sept. 1989: A1.

83. Meganack's speech was delivered by another village member at a conference of Oiled Mayors. It was reportedly ghost-written by a non-Native lawyer for the village, so its authenticity has come into question. However, the collective writing of such a document would not be unusual under these circumstances. Walter Meganack, "The Day the Water Died," in *The Day the Water Died,* 44.

84. Ibid., 45.

85. The media often made specific comparisons between Natives and otters. For example, *Newsweek* reported of Exxon's supplying of meat and other groceries to the villages, "Of course, the amount spent to feed the villagers was nothing compared with what was lavished on the otters." Jerry Adler, "Alaska after Exxon," *Newsweek,* 18 Sept, 1989: 50.

86. Rita Giordano, "Native Alaskans Sue Over Exxon Oil Spill," *Newsday,* 20 Apr. 1989: 15.

87. Egan, A1.

88. Ibid.

89. Ibid.

90. McClintock, 11.

91. Arthur N. Gilbert and Michael Barkun, "Disaster and Sexuality," *Journal of Sex Research* 17 (1982): 288–99.

92. In a letter to the *New York Times* in Apr. 1995, the Unabomber claimed that he had killed Thomas Mosser because he worked at a public relations firm that had helped restore Exxon's image after the spill.

93. Kirkpatrick Sale, "Toward a Portrait of the Unabomber," *New York Times* 6 Aug. 1995: sec. 4, p. 15.

94. Don Ihde, *Technology and the Lifeworld: From Garden to Earth* (Bloomington: Indiana University Press, 1990), 20.

95. Joseph G. Jorgensen, "Ethnicity, Not Culture? Obfuscating Social Science in the *Exxon Valdez* Oil Spill Case," *American Indian Culture and Research Journal* 19 (Fall 1995): 1–124.

96. In a letter defending himself against Jorgensen's accusations, Judge H. Russel Holland denied that these arguments over what does or does not constitute culture had any influence on his decision to refuse the claim that a way of life had been harmed. He argued that this was a strict decision based on legal constraints, and was not influenced by Exxon consultant Paul Bohannan. Jorgensen replied with a further attack on Holland's assessments of Native culture. See H. Russel Holland, Letter to the Editor, *American Indian Culture and Research Journal* 20 (Fall 1996): 167–70; Jorgensen, "'Ethnicity, Not Culture? . . .' A Reply," *American Indian Culture and Research Journal* 20 (Fall 1996): 171–75. The case was further complicated by Exxon and the court's refusal to accept nonmarket valuations of damage. Because people in subsistence economies do not keep written records of their catches or value them in cash terms, their damage claims were denied unless they could provide such evidence. See John Duffield,

"Nonmarket Valuation and the Courts: The Case of the *Exxon Valdez*," *Contemporary Economic Policy* 15 (Oct. 1997): 98–110.

97. Allan Young, *The Harmony of Illusions: Inventing Post-Traumatic Stress Disorder* (Princeton, N.J.: Princeton University Press, 1995), 127.

98. Impact Assessment, Inc., *Final Report: Economic, Social, and Psychological Impact Assessment of the* Exxon Valdez *Oil Spill* (La Jolla, Calif.: Impact Assessment, 1990), x.

99. Terry Wilson, "Psychologist Says Stress Is Byproduct of Oil Spill Cleanup," *Valdez Vanguard,* 25 Oct. 1989: A1, A6. Spill stress video, outtakes, videographer Tom McDowell, Homer, 19 July 1989.

100. Beth Tornes, "'The Sound Is Still Not Clean': The Impact of the *Exxon Valdez* Oil," *News from Indian Country,* 9 (Sept. 1995): 26.

101. Rose A. Horowitz, "Lawsuit Claimed Hazelwood Abusive," *Journal of Commerce* 5 Apr. 1989: A1.

102. David Lebedoff, *Cleaning Up: The Story behind the Biggest Legal Bonanza of Our Time* (New York: Free Press, 1997), 176.

103. Jorgensen, 39.

104. Lawrence Palinkas et al., "Social, Cultural, and Psychological Impacts of the *Exxon Valdez* Oil Spill," *Human Organization* 52 (Spring 1993): 10.

105. Christopher Wooley, who assessed damage to archeological sites after the spill, has made the same argument: "The Alutiiq people were not propelled from the Stone Age into contemporary American life by the oil spill." "Alutiiq Culture before and after the *Exxon Valdez* Oil Spill," *American Indian Culture and Research Journal* 19 (Fall 1995): 140. Furthermore, according to reports from villages themselves, many inhabitants used money from cleanup work to buy snow machines, TVs, satellite dishes, appliances, furniture, and diesel heaters.

106. Christopher L. Dyer, Duane A. Gill, and J. Steven Picou, "Social Disruption and the Valdez Oil Spill: Alaskan Natives in a Natural Resource Community," *Sociological Spectrum* 12 (1992): 120. See also Christopher L. Dyer, "Tradition Loss as Secondary Disaster: Long-Term Cultural Impacts of the *Exxon Valdez* Oil Spill," *Sociological Spectrum* 13 (1993): 65–88. In 1996, Gill and Picou published a textbook version of their work in which they included many more actual Native voices, drawing from second-hand sources, mostly from Meganack's speech. See J. Steven Picou, Duane A. Gill, and Maurie J. Cohen, eds., *Technological Disaster at Valdez: Readings on a Modern Social Problem* (Dubuque, Iowa: Kendall-Hunt, 1996).

107. McClintock, 2.

108. Ibid., 54.

109. Ibid., 24.

110. Ibid., 26.

111. Jeff Wheelwright, "Exxon Was Right. Alas," *New York Times,* 31 July 1994: lat. ed., sec. 4, p. 15.

112. Wheelwright, 13–14.

113. Piper, 111.

114. Wheelwright, 275.

115. L. G. Rawl and L. R. Raymond, "Letter to the Shareholders," *1989 Annual Report* (New York: Exxon, 1989), 2–3.

116. R. W. Scott, "Burn It; *Exxon Valdez* Oil Spill," *World Oil,* 208 (May 1989): 5.

117. Wheelwright, 271.

118. Kai Erikson, *A New Species of Trouble: Explorations in Disaster, Trauma, and Community* (New York: Norton, 1994), 241.

119. Jorgensen, 32.

120. Qtd. in Tim Bristol, "Hard Choices for a Shaky Future: Alaska Natives in the Wake of the *Exxon Valdez*," *Native Americans* (Spring–Summer 1995): 41.

121. Qtd. in Tornes, 26.

122. Megan McAtee, "*Exxon Valdez* Funds Target Subsistence Concerns," *Tundra Times,* 16 Aug. 1995: 5. I am also grateful to Martha Vlasoff of the EVOS Trustee Council for calling my attention to Tom Albert's bowhead whale studies that incorporate traditional knowledge.

123. EVOS Trustees, "Restoration Framework," Apr. 1992, in Piper, 171.

124. The EVOS Trustee Council, which funds much of the research being done on wildlife populations in the sound, reported in 1996 that although some marine mammals, birds, and fish had recovered or were making progress toward recovery, "Harbor seals . . . continue to decline at a rate of about six percent per year. Certain oiled seabird colonies have not recovered. . . . A much studied pod of killer whales in Prince William Sound has suffered additional losses in the past two years, and the social structure of the pod seems to be disintegrating. Residents in the spill area continue to deal with disruptions to commercial fishing and their subsistence way of life." Furthermore, oil, in the form of asphalt and tar, can still be found on the beaches, especially on the islands hardest hit. The council was formed to disperse funds from the $900-million settlement awarded to the State of Alaska. EVOS Trustee Council, *1996 Status Report* (Anchorage: Oil Spill Information Center, 1996), 5.

125. Del Jones, "'Fortune' 500 Profits Up 15%," *USA Today,* 29 Mar. 1994: 2.

126. "Executive Compensation: Exxon CEO Got 17% Raise in Salary, Bonus," *Atlanta Journal and Constitution,* 22 Mar. 1997: E7.

127. William J. Small, "*Exxon Valdez:* How to Spend Billions and Still Get a Black Eye," *Public Relations Review* 17 (1991): 9. See also Patrick Daley, "Sad Is Too Mild a Word: Press Coverage of the *Exxon Valdez* Oil Spill," *Journal of Communication* 41 (Autumn 1991): 42–58; David E. Williams and Bolanle A. Olaniran, "Exxon's Decision-Making Flaws: The Hypervigilant Response to the *Valdez* Grounding," *Public Relations Review* 20 (1994): 5–19; William Benoit, "Image Repair Discourse and Crisis Communication," *Public Relations Review* 23 (1997): 177–87; Phillip Perry, "Exxon Falters in PR Effort Following Alaskan Oil Spill," *O'Dwyer's PR Services Report,* July 1989: 1; David E. Williams and Glenda Treadway, "Exxon and the Valdez Accident: A Failure in Crisis Communication," *Communication Studies* 43 (Spring 1992): 56–65; Lee W. Baker, *The Credibility Factor: Putting Ethics to Work in Public Relations* (Homewood, Ill.: Business One Irwin, 1993), 38–45.

128. James E. Lukaszewski, "The *Exxon Valdez* Paradox," in *Crisis Response: Inside Stories on Managing Image under Siege,* ed. Jack Gottschalk (Detroit: Gale Research, 1993), 188, 208.

129. Ron Rogers, "Anatomy of a Crisis," *Crisis Response,* 124.

130. Lukaszewski, 209.

131. P. Sherman Stratford, "Smart Ways to Handle the Press," *Fortune,* 19 June 1989: 69.

132. Patty Tascarella, "Dancing at the Top; Burson Marsteller Managing Director Laura Gongos," *Pittsburgh Business Times,* 21 July 1997: 8.

133. Stuart Ewen, *PR! A Social History of Spin* (New York: Basic, 1996), 6.

134. Joe Nichols, "Oil Spills, Media Pressure, and Environmental Resilience," *Ocean Orbit: Newsletter of the International Tanker Owners Pollution Federation*, Feb. 1992: 7.

135. Lawrence Rawl, Interview, *Fortune*, 8 May 1989: 8.

136. Lawrence Rawl, Interview, *Time*, 26 Mar. 1990: 62.

137. Ibid., 63.

138. Qtd. in Adler, 54.

139. "A Total Response: Long Hours, Strong Morale Highlight Effort in Alaska," *Profile* 28 (May–June 1989): 13.

140. Phillip M. Perry, "Exxon Falters in PR Effort Following Alaskan Oil Spill," *O'Dwyer's PR Services Report*, July 1989: 1.

141. Ortega, 13.

142. Ibid.

143. Ibid.

144. Tom Crosser, qtd. in George Frost, "Veco Reduces Cleanup Jobs," *Anchorage Daily News*, 30 Apr. 1989: B-1.

145. Jay Mathews, "*Anchorage Times* Wages Unusual Paper War," *Washington Post*, 31 Mar. 1990: A6. Allen wanted the *Anchorage Times* to rival the *Anchorage Daily News*, which he perceived as being liberal and anti-oil.

146. Exxon, *Progress in Alaska* (video), 1989.

147. Norman Soloman, qtd. in George Frost, A4.

148. Personal interview with Valdez Museum director Joe Leahy, 1 Aug. 1996. See also City of Valdez and Alaska Humanities Forum, *Sea of Oil*, Affinity Films, 1990.

149. Exxon, *Progress in Alaska*, 18 May 1989.

150. See "Progress Marked in Alaska as Cleanup Prepares to 'Winterize,'" *Profile*, Aug. 1989: 9; "A Total Response: Long Hours, Strong Morale Highlight Effort in Alaska," *Profile*, May–June 1989: 4–13.

151. Charles Wohlforth, "Season of the Spill: A Reporter Reflects," in *Cries from the Heart*, ed. Jan O'Meara (Homer, Alaska: Wizard Works, 1989), 63–68.

152. *Sea of Oil*.

153. Keith Schneider, "That Sinking Feeling," *Relate*, 24 Apr. 1989: 25.

154. Wohlforth, 65.

155. Peter Nulty and Frederick H. Katayama, "Corporate Performance: Exxon's Problem: Not What You Think," *Fortune*, 23 Apr. 1990: 202.

156. Michael Satchell and Betsey Carpenter, "A Disaster That Wasn't," *U.S. News & World Report*, 18 Sept. 1989: 62.

157. Nina Munk, "'We're Partying Hearty,'" *Forbes*, 24 Oct. 1994: 84.

158. Although Exxon's "globalization" extended to many aspects of the spill's science, a good example lies in the discussion of how much shoreline was oiled. As Wheelwright relates in his book, Al Maki declared on *Nightline* that only 10 percent of the "linear shoreline of Prince William Sound" was oiled. But as Wheelwright points out, the length of the sound's shoreline was open to interpretation, and Exxon's estimate was the highest in use (98–99). Similarly, Exxon scientists framed the loss of wildlife in terms of global populations. By expanding frames of reference, Exxon could diminish spill impacts.

159. Stephanie Pain, "The Two Faces of the Exxon Disaster," *New Scientist*, 22 May 1993: 11.

160. Paul Shrivastava, *Bhopal: Anatomy of a Crisis* (Cambridge, Mass.: Ballinger, 1987), 89.

161. George Frost, A4.

162. Jenifer M. Baker, Robert B. Clark, and Paul F. Kingston, *Environmental Recovery in Prince William Sound and the Gulf of Alaska* (Edinburgh: Institute of Offshore Engineering, 1990), 5.

163. Ibid., 3.

164. Jenifer M. Baker, Robert B. Clark, Paul F. Kingston, *Two Years after the Spill: Environmental Recovery in Prince William Sound and the Gulf of Alaska: November 1991 Supplement* (Edinburgh: Institute of Offshore Engineering, 1991), 8.

165. Exxon, *The Abundant Bald Eagles of Prince William Sound, Alaska* (Houston: Exxon, 1991); Exxon, *Sea Otters Thrive in Prince William Sound, Alaska* (Houston: Exxon, 1991).

166. See, for example, Exxon's advertisement, "Pictures of Health," *Alaska,* Jan. 1992: back cover.

167. Exxon, *Alaska Update,* 18 Sept. 1990.

168. Exxon, *Annual Report,* (New York: Exxon, 1989), 5.

169. Alan Maki, "Perspective: Ecological Recovery Following the Oil Spill," Michigan State University, 24 Mar. 1997.

170. Keith Michael Hearit argues that postmodern corporations facing crisis may turn to a "transcendent apologia," a "strategy in which guilt is avoided through its redefinition into a virtue" by shifting the debate to a broader context. In "On the Use of Transcendence as an Apologia: A Strategy: The Case of Johnson Controls and its Fetal Protection Policy," *Public Relations Review* 22 (1997): 217–32.

171. *Scientists and the Alaska Oil Spill* (video), Exxon, 1989.

172. Edward Bernays's term for public relations.

173. From a list of species "not recovering," in the *Exxon Valdez* Oil Spill Trustee Council's 1996 Status Report.

Chapter 5: The Bhopal Effect

1. Qtd. in Bharat Desai, "Bhopal Gas Tragedy: A Mockery of Compensation," *India Today,* 15 Mar. 1997: 27.

2. Upendra Baxi, Introduction, *Valiant Victims and Lethal Litigation: The Bhopal Case,* ed. Upendra Baxi and Amita Dhanda (Bombay: Tripathi, 1990), liv.

3. Elaine Scarry, *The Body in Pain* (New York: Oxford University Press, 1985), 5.

4. The death figures from Bhopal are highly disputed, and may have reached as high as 10,000 immediate deaths. For a discussion of the court's controversial casualty figures, see Claude Alvares, "Bhopal Ten Years After," in *Bhopal: The Inside Story, Carbider Workers Speak Out on the World's Worst Industrial Disaster,* ed. T. R. Chouhan (New York: Apex, 1994), 138–42.

5. Ashis Gupta, *Ecological Nightmares and the Management Dilemma: The Case of Bhopal* (Delhi: Ajanta, 1991), 26; Lalit Shastri, *Bhopal Disaster: An Eyewitness Account* (New Delhi: Criterion, 1985), 72–73.

6. S. R. Deshmukh, Statement, in *Bhopal: The Inside Story,* 94.

7. Veena Das, *Critical Events: An Anthropological Perspective on Contemporary India* (Delhi: Oxford University Press, 1995), 19.

8. Ibid., 174.

9. For an excellent account of how U.S. and Indian courts protected the interests of Union Carbide and the Indian government at the expense of the victims, see Jamie Cas-

sels, *The Uncertain Promise of Law: Lessons from Bhopal* (Toronto: University of Toronto Press, 1993).

10. Cassels, 289–93; Desai, 27; Indira Jaising, "Legal Let-Down," in *Bhopal: The Inside Story,* 181–212; Praful Bidwai, "Bhopal: A Dismal Balance Sheet," in *Bhopal: The Inside Story,* 168–70; Armin Rosencranz, Shyam Divan, and Antony Scott, "Legal and Political Repercussions in India," in *Learning from Disaster: Risk Management after Bhopal,* ed. Sheila Jasanoff (Philadelphia: University of Pennsylvania Press, 1994), 44–65; Sanjay Kumar, "Bhopal Claims Are Genuine," *New Scientist,* 5 Feb. 1994: 9; Molly Moore, "The Second Disaster in Bhopal," *Business and Society Review,* Jan. 1994: 26.

11. Lee Wilkins discusses the invasive use of photography at Bhopal, which she says both presented stereotypical views of Indians and gave images of all humanity's vulnerability to such catastrophes, in Wilkins, *Shared Vulnerability: The Media and American Perceptions of the Bhopal Disaster* (New York: Greenwood, 1987), 135–39. Suketu Mehta puts the Bhopal photographs in the context of a "pornography of images of disaster in the Third World" in "Bhopal Lives: The 1984 Union Carbide Toxic-Gas Disaster Killed 10,000 People—and Has Changed Everything for Its Survivors," *Village Voice,* 3 Dec. 1996: 50. In their discussion of the media's globalizing images of violence, especially from the Third World, Arthur Kleinman and Joan Kleinman suggest that suffering has become commercialized and exchanged for profit, desensitizing the viewer. See "The Appeal of Experience; the Dismay of Images: Cultural Appropriations of Suffering in Our Times," *Daedalus* 125 (Jan. 1996): 1–24. My own reading of the Bhopal photographs is also inspired by Cathy N. Davidson, "Photographs of the Dead: Sherman, Daguerre, Hawthorne," *The South Atlantic Quarterly* 89 (Fall 1990): 667–701; Roland Barthes, *Camera Lucida: Reflections on Photography,* trans. Richard Howard (New York: Hill & Wang, 1981); and Susan Sontag, *On Photography* (New York: Farrar, Straus & Giroux, 1977).

12. The *India Today* image is credited to the Delhi Forum, Suara Sam, Malaysia, 1986, and appears with Ashok Bhargava's "The Bhopal Incident and Union Carbide: Ramifications of an Industrial Accident," *Bulletin of Concerned Asian Scholars* 18 (Oct.–Dec. 1986): 2–18.

13. Ramesh, "Memories: The Unforgettable Night," *No More Bhopals,* 1997, available online at http://www.ucaqld.com.au/community/bhopal/ (27 Feb. 1998).

14. *Bhopal: The Inside Story,* 39.

15. Qtd. in Tim McGirk, "Indian Victims' Anger Lives On," *Independent,* 9 Dec. 1991: 14.

16. Abdul Jabbar Khan, Interview, *Bulletin of Concerned Asian Scholars* 26 (Jan.–June 1994): 14.

17. P. S. Chauhan, *Bhopal Tragedy: Socio-Legal Implications* (New Delhi: Rawat, 1996), 60.

18. Brojendra Nath Banerjee, *Environmental Pollution and the Bhopal Killings* (Delhi: Gian, 1987), 36.

19. Chauhan, 50–53.

20. Ira Chernus claims that a worshipful attitude toward nuclear weapons has resulted in a new religion that replaced God with a numinous technological force in *Dr. Strangegod: On the Symbolic Meaning of Nuclear Weapons* (Columbia: University of South Carolina Press, 1986).

21. Robert D. McFadden, "India Disaster: Chronicle of a Nightmare," *New York Times,* 10 Dec. 1984: A1.

22. Pico Iyer, "India's Night of Death," *Time,* 17 Dec. 1984: 23; Dean Brelis, "'I Thought I Had Seen Everything,'" *Time,* 17 Dec. 1984: 25.

23. Mark Whitaker, "'It Was Like Breathing Fire,'" *Newsweek,* 17 Dec. 1984: 26.

24. Dan Kurzman, *A Killing Wind: Inside Union Carbide and the Bhopal Catastrophe* (New York: McGraw-Hill, 1987), 1.

25. Ibid., 92.

26. Ibid., 59.

27. Qtd. in John Elliott, "Fear of Long-Term Illness in Bhopal," *Financial Times,* 17 Jan. 1985: 4.

28. A socialist organization, once led by nuclear physicist Frederic Joliot-Curie, based in Helsinki, often the object of U.S. right-wing paranoia about communist infiltration.

29. The most complete rendering of this hypothesis can be found in Brojendra Nath Banerjee, *Bhopal: Accident or Experiment* (Delhi: Paribus, 1986).

30. N. Rajan, "Carbide Lobby Active in Bhopal?" in *Bhopal: From Hiroshima to Eternity,* ed. S. B. Kolpe (Bombay: Kolpe, 1985), 9–16.

31. Gupta, 14–15.

32. Russell Miller, "Victims the World Forgot," *Mail on Sunday* (London), 32 Mar. 1997: 10.

33. Sanjoy Hazarika, *Bhopal: The Lessons of a Tragedy* (New York: Penguin, 1987), 97–99; Anees Chishti, *Dateline Bhopal: A Newsman's Diary of the Gas Disaster* (New Delhi: Concept, 1986), 48; P. S. Vivek, *The Struggle of Man against Power: Revelation of 1984 Bhopal Tragedy* (Delhi: Himalaya, 1990), 40–41.

34. Michael R. Reich, "Toxic Politics and Pollution Victims in the Third World," *Learning from Disaster,* 187–88.

35. Jabbar Khan, "A Call for Action," *Bulletin of Concerned Asian Scholars* 26 (Jan.–June 1994): 19.

36. Shahazadi Bahar, Interview, *Just Cause* 1 (Nov. 1994): 13.

37. Amrita Basu, "Bhopal Revisited: The View from Below," *Bulletin of Concerned Asian Scholars* 26 (Jan.–June 1994): 3–14. Radha Kumar puts the Bhopal women's activism in the larger context of Indian women's political movements in *The History of Doing: An Illustrated Account of Movements for Women's Rights and Feminism in India, 1800–1990* (New York: Verso, 1993), 187–90.

38. Chishti, 52.

39. For a summary of findings on women's health, see Ingrid Eckerman's final report on Bhopal women and children survivors' health for the International Medical Commission, 19 Aug. 1996. She also reports on social conditions for women, including the ostracizing and lack of marriageability of gas-affected girls, available online at http://www.ucaqld.com.au/community/bhopal/imcb/report.html (25 Sept. 1998).

40. C. Sathyamala, "The Condition of Bhopal's Women," in *Bhopal: The Inside Story,* 130.

41. Banerjee, *Environmental Pollution,* 41–45; Shastri, 39.

42. Akeeza Bi, qtd. in Chouhan, Preface, *Bhopal: The Inside Story,* 7.

43. Sheila Tefft, "Cloud of Fear Still Hangs over Bhopal," *Chicago Tribune,* 1 Dec. 1985: C8. See also William K. Stevens, "Gas Leak Death Rate Dips; More Hopeful Outlook Seen," *New York Times,* 9 Dec. 1984: 22; and Rajiv Desai, "An Ill-Wind for the Once Prosperous People of Bhopal," *Chicago Tribune,* 30 Nov. 1986: C14.

44. Jabbar Khan, Interview, 14.

45. Susanna Davky, "Bhopal Disaster: A Personal View," in *Bhopal: From Hiroshima to Eternity,* 25; see also Kurzman, 153.

46. Edward Bowman and Howard Kunreuther, "Post-Bhopal Behavior at a Chemical Company," *Journal of Management Studies* 25 (1988): 391, 393.

47. For analyses of the news presentation of Union Carbide, see Michael J. Lynch,

Mahesh K. Nalla, and Keith W. Miller, "Cross-Cultural Perceptions of Deviance," *Journal of Research in Crime and Delinquency* 26 (1989): 7–35; Wilkins, 15–19, 81–92.

48. Thierry C. Pauchant and Ian I. Mitroff, *Transforming the Crisis-Prone Organization: Preventing Individual, Organizational, and Environmental Tragedy* (San Francisco: Jossey-Bass, 1992), 8, 15.

49. Qtd. in Judith H. Dobrzynski et al., "Union Carbide Fights for Its Life," *Business Week,* 24 Dec. 1984: 52.

50. Richard I. Kirkland, Jr., "Union Carbide: Coping with Catastrophe," *Fortune,* 7 Jan. 1985: 52.

51. Michael Silva and Terry McGann, *Overdrive: Managing in Crisis-Filled Times* (New York: Wiley, 1995), 92–93.

52. Charles P. Alexander, "A Calamity for Union Carbide: The Financial Future of the Chemical Giant Is in Question," *Time,* 17 Dec. 1984: 38.

53. Jesse Werner, Letter to the editor, *Chemical Week,* 15 Jan. 1986: 4. Under Werner's successor, Samuel Heyman, GAF made a hostile bid for Union Carbide because it was perceived as being in potential financial trouble because of Bhopal. The bid failed.

54. Qtd. in Alexander Cockburn, "Flacks: Public Relations Specialists," *Playboy* 34 (Jan. 1987): 98.

55. According to an Indian consul, Anderson's visit was considered a breach of protocol, and his offer to meet with Gandhi was arrogance and "macho posturing." See W. David Gibson, "How the Indians View Carbide," *Chemical Week,* 18 Dec. 1985: 12.

56. Upendra Baxi explains Union Carbide's corporate tactics and India's formulation of a standard of justice for multinational corporations. In Baxi, Preface, *Valiant Victims,* i–lxix. For a full discussion of the legal negotiations resulting from the Bhopal disaster, see Cassels.

57. Terry Dodson, "Spectre of Bhopal Haunts a Troubled Union Carbide," *Financial Times,* 13 Aug. 1985: 4.

58. Joani Nelson-Horchler, "Fallout from Bhopal: Industry Confronts the Long-Term Consequences," *Industry Week,* 13 May 1985: 44.

59. Judith H. Dobrzynski, "Morton Thiokol: Reflections on the Shuttle Disaster," *Business Week,* 14 Mar. 1988: 86.

60. Stuart Diamond, "Warren Anderson: A Public Crisis, A Personal Ordeal," *New York Times,* 19 May 1985: sec. 3, p. 1.

61. W. David Gibson, "Union Carbide Restructuring under Stress," *Chemical Week,* 27 Nov. 1985: 94.

62. Kirkland, 50.

63. Clemens P. Work, "Inside Story of Union Carbide's India Nightmare," *U.S. News & World Report,* 21 Jan. 1985: 51.

64. Dobrzynski, "Union Carbide Fights," 53.

65. Peter Stoler, "Frightening Findings at Bhopal; Union Carbide and India Begin to Uncover What Happened," *Time,* 18 Feb. 1985: 78. See also Thomas J. Lueck, "Uneasy Headquarters in Carbide Accident," *New York Times,* 16 Dec. 1984: 11CN, 1; Thomas J. Lueck, "Crisis Management at Carbide," *New York Times,* 14 Dec. 1984: D1; N. R. Kleinfield, "When Scandal Haunts the Corridors," *New York Times,* 7 July 1985: sec. 3, p. 1.

66. "Credibility, Planning Keys to Crisis Communication," *Occupational Hazards* 58 (May 1996): 98.

67. Warren Anderson, Letter, *Union Carbide World,* Jan.–Feb. 1985: cover.

68. As Anderson told a *New York Times* reporter, "Union Carbide is very much like a family where when there is a problem you close ranks." In Diamond, sec. 3, p. 1.

69. Ron Van Mynen, qtd. in Gregg LaBar, "Citizen Carbide? The Stigma of the Bhopal Tragedy Still Haunts Union Carbide Chemicals and Plastics Company, Inc.," *Occupational Hazards* 53 (Nov. 1991): 33.

70. Marilyn A. Harris, "For Carbide's U.S. Workers, Lost Faith—And Lost Jobs," *Business Week,* 25 Nov. 1985: 98.

71. Warren M. Anderson, Letter to the editor, *Business Week,* 20 Jan. 1986: 20.

72. Chouhan, Preface, *Bhopal: The Inside Story,* 11–12; "Carbide's Sabotage Theory," *Bhopal: The Inside Story,* 61–70 (Verma's own statement appears on pp. 85–89); Desai, 14; Russell Miller, "Victims the World Forgot," *Mail on Sunday* (London), 23 Mar. 1997: 10; W. Lepkowski, "Union Carbide Presses Bhopal Sabotage Theory," *Chemical and Engineering News* 66 (July 1988): 8–11; Langdon Brockinton and Donald P. Burke, "Why Carbide Says It Was 'A Deliberate Act' at Bhopal," *Chemical Week,* 26 Nov. 1986: 8; Laurie A. Rich, "Carbide Realleges Sabotage," *Chemical Week,* 20 Aug. 1986: 13.

73. Harris, 98; Wil Lepkowski, "The Restructuring of Union Carbide," in *Learning from Disaster,* 25–26.

74. Kleinfield, 1.

75. "Carbide Closes Bhopal Pesticide Plant," *Washington Post,* 12 July 1985: B1.

76. Bowman and Kunreuther, 397.

77. After a government study noted that the Three Mile Island, Bhopal, and *Challenger* accidents had all occurred in the wee hours of the morning, sleep researchers began to note the importance of circadian rhythms to industrial safety, and a lucrative consulting business began to design workers' sleep and work patterns. See U.S. Office of Technology Assessment, *Biological Rhythms: Implications for the Worker* (Washington, D.C.: U.S. Government Printing Office, 1992).

78. Barbara Ettorre, "The Care and Feeding of a Corporate Reputation," *Management Review* 85 (June 1996): 39.

79. Eileen Murray and Saundra Shohen, "Lessons from the Tylenol Tragedy on Surviving a Corporate Crisis," *Medical Marketing & Media* 27 (Feb. 1992): 14.

80. Norman R. Augustine, "Business Crises: Guaranteed Preventatives—And What to Do After They Fail," *Executive Speeches* 9 (June–July 1995): 28.

81. Leon H. Mayhew, *The New Public: Professional Communication and the Means of Social Influence* (New York: Cambridge University Press, 1997), 236–37.

82. Jennifer Nash and John Ehrenfeld, "Code Green: Business Adopts Voluntary Environmental Standards," *Environment* 38 (Jan. 1996): 16–20, 36–45; Ronald Begley, "After Bhopal: A CMA 'War Room,' New Programs, and Changed Habits," *Chemical Week,* 7 Dec. 1994: 32.

83. David Dembo, *Abuse of Power: Social Performance of Multinational Corporations: The Case of Union Carbide* (New York: New Horizons, 1990), 75; "We All Live in Bhopal," South Downs Earth First!, http://www.hrc.wmin.ac.uk/campaigns/ef/dt/bhopal.html (19 Feb. 1998).

84. David Hunter, "CAPS and Responsible Care," *Chemical Week,* 9 July 1997: 5.

85. See, for example, William J. Benoit, *Accounts, Excuses, and Apologies: A Theory of Image Restoration Strategies* (Albany, N.Y.: SUNY University Press, 1995).

86. Shelley A. Hearne, "Tracking Toxics: Chemical Use and the Public's 'Right-to-Know,'" *Environment* 38 (July–Aug. 1996): 4–9, 28–34.

87. Melinda Beck, "Could It Happen in America?" *Newsweek,* 17 Dec. 1984: 38; Jamie Murphy, "Could It Happen in West Virginia?" *Time,* 17 Dec. 1984: 36. See also Michael Brown's description of Institute residents' response to Bhopal in Brown, *The Toxic Cloud* (New York: Harper & Row, 1987), 223–24.

88. Hans Jonas, *The Imperative of Responsibility: In Search of an Ethics for the Technological Age* (Chicago: University of Chicago Press, 1984), 26–27.

89. Ibid., 30.

90. Thomas J. Lueck, "Company Defends Plant Site in India," *New York Times,* 6 Dec. 1984: A1.

91. Edwin D. Hoffman, Remarks, Town Meeting of Institute-Dunbar Citizens, Shawnee Community Center, Dunbar, 9 Dec. 1984. Included in People Concerned about MIC, *From Bhopal to Institute,* typescript brochure, c. 1986.

92. Paul Nuchims, Personal interview, 22 Mar. 1998.

93. People Concerned about MIC, Letter to Mayor of Bhopal, 21 Mar. 1985.

94. Paul Nuchims, Letter to Warren Anderson, 8 Apr. 1985.

95. Casey Bukro, "Safety Issue Hits Home in Chemical Valley," *Chicago Tribune,* 18 Aug. 1985: 5.

96. "Carbide Plant Leaks, 150 Ill," *Chicago Tribune,* 12 Aug. 1985: C1.

97. Casey Bukro, "2d Carbide Plant Leaks Chemical," *Chicago Tribune,* 14 Aug. 1985: C1. According to the EPA's TRI for 1998 (the last report available as of this writing), Carbide's South Charleston plant reported releases of hundreds of thousands of pounds of toxins, including acetaldehyde and ethylene oxide, both known carcinogens. Union Carbide may have rid itself of the highly visible Institute plant, but its South Charleston plant still ranks among the top ten chemical polluters in West Virginia, along with Rhone Poulenc's.

98. *Chemical Valley,* dir. Mimi Pickering and Anne Lewis, Appalshop, 1991.

99. The group's new charter, after the aldicarb leak, appears in its newsletter, *Downwind* 1 (Spring 1986): 5.

100. Elisabeth Kirschner, "Risk Management: Beyond Assessment," *Chemical Week,* 26 Jan. 1994: 4.

101. Mildred S. Myers and S. Lee Jerrell, "Dealing with an 'Irrational' Public," *Across the Board* 34 (Jan. 1997): 39.

102. Kara Sissell, "Worst-Case Scenarios Test Responsible Care," *Chemical Week,* 2 July 1997: 3.

103. "Coming Together as a Community," *Public Management* 76 (Oct. 1994): 24.

104. The Safety Street organizers produced a large packet of information on which much of my analysis is based: Kanawha/Putnam Local Emergency Planning Committee, *Safety Street, Managing Our Risks Together: A Guide to the Kanawha Valley Hazard Assessment Project,* 4 June 1994.

105. Ken Ward, Jr., "Displays Downplay Concerns," *Charleston Gazette,* 5 June 1994: 1A, 4A.

106. Joyce Carol Oates, *You Must Remember This* (New York: Dutton, 1987), 22.

107. Ken Ward, Jr., "Worst-Case Scenarios Won't Be the Worst," *Charleston Gazette,* 27 Apr. 1994: 2A.

108. Paul Nuchims, "A Short History of Sheltering in Place," delivered at the National Institute for Chemical Studies' conference, Protecting the Public: Protective Actions During Chemical Emergencies, 21–22 Sept. 1995, Charleston, W.V.

109. A videoscript of Emergency Warning System, produced by the Kanawha/Putnam Local Emergency Planning Committee, is available in the *Safety Street* guide.

110. James Brandt, "A Community Pulling Together," in *Safety Street* guide.

111. Ryan McGinn Samples, *'Safety Street' Post Test Survey* (Charleston, W.V.: Ryan McGinn Samples, 1994).

112. Ken Ward, Jr., "What to Do? Latest Chemical Leak Raises Same Questions on Health, Notification," *Charleston Gazette,* 26 Feb. 1996: C1.

113. Ken Ward, Jr., "Activist Blasts Chemical Companies for Downplaying Recent Plant Leak," *Charleston Gazette,* 26 Jan. 1995: 7A.

114. "Environmentalists Honored for Their Struggle," *Charleston Daily Mail,* 22 Feb. 1995: 6A.

115. Ken Ward, Jr., "Environmental Activist Finds She Has Rare Disease," *Charleston Gazette,* 22 Feb. 1995: 1C.

116. Ibid.

Index

Ann Larabee, an associate professor of American thought and language at Michigan State University, has published numerous articles on technological disaster. Her work on the *Challenger* disaster was awarded *Postmodern Culture Journal*'s Electronic Text Award. She has served as president of the Lansing Area AIDS Network.

Typeset in 10.5/13 Adobe Minion
Designed by Copenhaver Cumpston
Composed by Celia Shapland
for the University of Illinois Press

University of Illinois Press
1325 South Oak Street
Champaign, Il 61820-6903
www.press.uillinois.edu